Acoustic Metamaterials

Sz-Chin Steven Lin

Acoustic Metamaterials

Tunable Gradient-Index Phononic Crystals for Acoustic Wave Manipulation

LAP LAMBERT Academic Publishing

Impressum/Imprint (nur für Deutschland/only for Germany)
Bibliografische Information der Deutschen Nationalbibliothek: Die Deutsche Nationalbibliothek verzeichnet diese Publikation in der Deutschen Nationalbibliografie; detaillierte bibliografische Daten sind im Internet über http://dnb.d-nb.de abrufbar.

Coverbild: www.ingimage.com

Verlag: LAP LAMBERT Academic Publishing GmbH & Co. KG
Heinrich-Böcking-Str. 6-8, 66121 Saarbrücken, Deutschland
Telefon +49 681 3720-310, Telefax +49 681 3720-3109
Email: info@lap-publishing.com

Approved by: University Park, The Pennsylvania State University, Diss., 2012

Herstellung in Deutschland (siehe letzte Seite)
ISBN: 978-3-659-18654-7

Imprint (only for USA, GB)
Bibliographic information published by the Deutsche Nationalbibliothek: The Deutsche Nationalbibliothek lists this publication in the Deutsche Nationalbibliografie; detailed bibliographic data are available in the Internet at http://dnb.d-nb.de.

Cover image: www.ingimage.com

Publisher: LAP LAMBERT Academic Publishing GmbH & Co. KG
Heinrich-Böcking-Str. 6-8, 66121 Saarbrücken, Germany
Phone +49 681 3720-310, Fax +49 681 3720-3109
Email: info@lap-publishing.com

Printed in the U.S.A.
Printed in the U.K. by (see last page)
ISBN: 978-3-659-18654-7

Table of Contents

Dedication

Dedicated to my beloved family and friends

Chapter 1

Motivation and Overview

1.1 Motivation

The transmission of a localized mechanical disturbance in a gaseous, liquid, or solid medium to other regions of the medium is known as elastic or acoustic wave propagation. Earthquakes, the spreading of ripples on a water surface, and the musical tones of strings are some of the most common elastic wave phenomena. Elastic waves have been extensively studied since the seventeenth century. They can propagate in an elastic medium in different modes, depending upon the medium properties and boundary conditions. For instance, bulk acoustic waves (BAW) propagate in the forms of longitudinal (compressional) and transverse (shear) vibrations in the interior of the medium. Surface acoustic waves (SAW), also known as Rayleigh waves, propagate along the free surface of the semi-infinite medium with wave motion localized within about one wavelength depth. Furthermore, elastic waves propagate in thin parallel-sided solid plates, and the waves at interface between two solids are called Lamb waves and Stoneley waves, respectively.

Inevitably, the outward propagating acoustic waves from a source encounter discontinuities in the material properties. In such situations, incident waves are scattered into reflected and/or transmitted waves at the interface separating two different media. The existence of the scattering from discontinuities affects the behavior of acoustic waves. Standing waves, for example, are formed if the reflected waves interfere constructively with the incident waves. These scattered waves complicate the problem of acoustic waves propagating in structures comprised of two or more materials that differ in properties, but at the same time broaden the acoustic wave application scope. For example, information such as size and position of a fetus in the womb can be obtained by collecting the scattered acoustic waves from a fetus to produce a three-dimensional image. Similarly, this "pulse-echo" diagnostic technique has been widely used in seismology, non-destructive evaluation (NDE), and underwater communication.

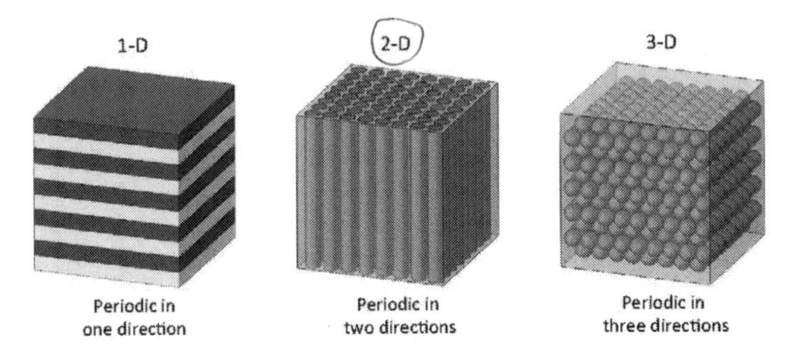

Figure 1.1. Schematics of one-, two-, and three-dimensional phononic crystals.

In the past decade, there has been a tremendous growth in interest in the two- and three-dimensional periodic structures due to their ability to manipulate the propagation of acoustic waves on the wavelength scale. These artificial periodic structures, coined as phononic crystals (PCs), are composed of arrays of elastic scatterers embedded in host materials of different elastic properties (Fig. 1.1). For wavelengths on the scale of a structure's periodicity, PCs exhibit *phononic band gaps*, the frequency ranges where acoustic waves in any vibration mode cannot penetrate the structure in any direction (Fig. 1.2). Phononic band gaps result from destructive interference of reflections from the periodic scatterers, making PCs promising candidates as perfect acoustic mirrors and acoustic filters of designated frequencies. In addition, any defect inside PCs will allow acoustic waves to vibrate only within the defect when operating within the phononic band gap, making them practical as acoustic resonators and acoustic waveguides [Fig. 1.3(a)].

Another great finding on PCs is *negative refraction*, the phenomenon where acoustic waves are refracted at an interface of materials in the reverse sense to that typically expected [Fig. 1.3(b)]. Negative refraction of acoustic waves does not exist between interfaces of natural materials. However, it has recently been theoretically and experimentally proven that PCs exhibit negative refraction at the edges of phononic band gaps. The

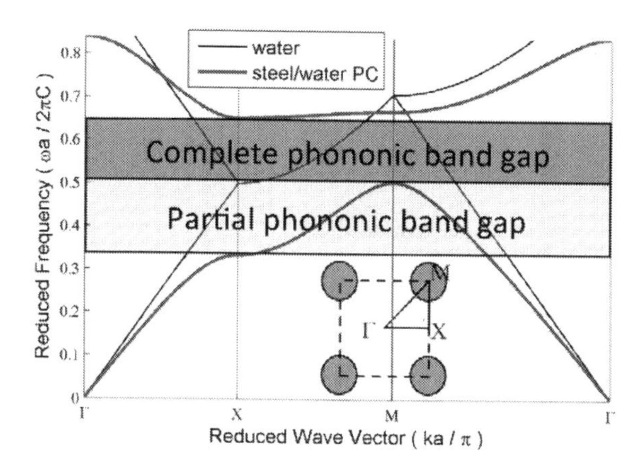

Figure 1.2. Band structure of water (black) and a steel/water PC (blue). The shaded areas denote the phononic band gaps.

strong anisotropy produced by the periodicity of the PCs is the cause of this extraordinary phenomenon. This discovery has paved the way for PC-based acoustic superlenses since it can overcome the diffraction limit. As a result, PCs are named acoustic metamaterials,

To this point, the investigation of PCs has been extended from BAW to SAW and Lamb waves, making PCs more compatible with state-of-the-art acoustic devices. Among these research, all existing PC functionalities take place within or above the first phononic band gap because at these frequency ranges PCs exhibit strong anisotropy to manipulate (block or bend) the propagation of acoustic waves. In these cases, the reduced frequency ($\Omega = fa/C$, where f is the working frequency, a is the periodicity of the scatterers, and C is the sound velocity in host material) is high. Since the reduced frequency, Ω, is proportional to the periodicity of the scatterers, a, a PC designed at higher bands may be too bulky to be integrated with other systems. For example, the period of a PC working at its first phononic band gap is in the order of meters, which makes it impossible to build noise-control earphones base on current PC technologies. In this regard, one part of this book is aimed toward developing a new class of compact PCs that is designed to work at the lowest band (below the first band gap) and demonstrate wave guiding functionalities

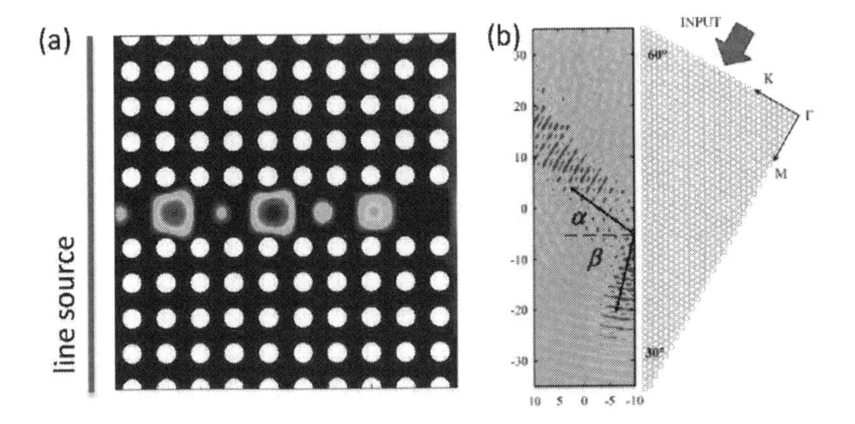

Figure 1.3. (a) An example of acoustic wave propagation in a two-dimensional PC waveguide. (b) Snapshot of the negative refraction experiment in a two-dimensional PC made of steel rods and immersed in water. (Source of image: doi: 10.1103/PhysRevB.77.014301)

similar to those of ordinary PCs. To achieve this aim, a novel approach to bend the propagation direction of acoustic waves by introducing the concept of gradient-index to PCs is proposed. In this book, we demonstrate large bending, focusing, and aperture modifying of acoustic beams can be conveniently achieved by gradient-index phononic crystals (GRIN PCs).

Another aspect of this book is about tuning the width and position of phononic band gaps of a PC, because all phenomena that PCs exhibit (including those by GRIN PCs) are more or less rely on phononic band gaps. While active tuning of the phononic band gaps is certainly desirable for future applications with enhanced functionalities, few attempts have been made to develop tunable PCs thus far. There is a strong need in developing a more practical method to actively tune the phononic band gaps. In this book, we present a comprehensive study of how phononic band gaps of a two-dimensional PC composed of anisotropic materials can be opened and closed simply by rotation of the cylinders. Our study offers a practical and effective waveguiding methodology for applications such as acoustic imaging, nondestructive evaluation (NDE), and lab-on-a-chip cell manipulations. Finally, when combining the two achievements in this book, we design a tunable GRIN PC for dynamic control of the propagation of acoustic waves.

1.2 Literature Review

Acoustic wave behavior in PCs is an emerging field that is considered to have great potential to enlarge acoustic wave applications. Its growth is partially induced by the successes of the rapid rising field of photonic crystals, which are periodic structures consist of two or more dielectric materials. The propagation of acoustic waves in one-dimensional PCs has been studied over fifty years, long before the periodic structures were called phononic crystals since the 1990s. Two-dimensional PCs were first studied by Sigalas and Economous [1] and Kushwaha *et al.* [2] in 1993. Their calculation of the acoustic band structure reveals the existence of phononic band gaps in two-dimensional artificial periodic structures. Since then, a great deal of work has been done in order to discover more functionality of PCs.

1.2.1 Theories

Over the past decade, the phononic band gaps of BAW in unbounded PCs, SAW in semi-infinite PCs and Lamb waves in PC plates have been extensively investigated through theoretical analyses and experimental demonstration [3, 4, 5, 6, 7, 8, 9, 10, 11, 12, 13, 14, 15, 16, 17]. On the theoretical side, a number of theories have been developed or adapted to analyze the acoustic wave propagation in PCs. The plane wave expansion (PWE) method [1, 2, 7, 10, 9, 14], multiple scattering theory (MST) [4, 18, 19, 20, 21, 15, 22], and finite-different time-domain (FDTD) method [23, 24, 25] are the three most popularly used tools among them. Based on the calculation mechanism, each method has its own strengths and weaknesses in dealing with periodic structures. The PWE method expends the elastic wave equation into Fourier series and solves the Eigenvalue problem. The PWE method is capable of analyzing unbounded or bounded PCs consisting of general anisotropic and piezoelectric materials [7, 10], but it is difficult to solve systems with complicated geometries and materials with large mismatches in acoustic impedance (e.g. fluid/solid systems). However, it is worth noting that the PWE method can be modified to approximate gaseous/solid systems [9]. On the contrary, MST is capable of analyzing structures with large elastic mismatch materials. It also has excellent calculation efficiency and thus is often used in complicated and three-dimensional PC analyses. Unfortunately, MST has not been improved to address systems containing anisotropic and piezoelectric materials. The FDTD method is widely used to simulate wave propagation. It converts the elastic wave equation to first-order difference equations and discretizes the simulation domain into grids. Tanaka *et al.* modified the FDTD method in 2000 to calculate the band structure of PCs [26]. It can virtually analyze PCs with any composition except piezoelectric materials, but suffers from massive calculation and numerical stability.

1.2.2 Experiments

On the experimental side, phononic band gaps of different acoustic wave modes have been verified in fluid/fluid, solid/fluid, fluid/solid, and solid/solid systems, soon after they were theoretically observed. For measuring phononic band gaps of BAW, PCs are normally fabricated in macroscopic scale for easier wave generation and detection using piezoelectric transducers. For SAW and Lamb waves, transmission through PCs is commonly detected by a scanning laser interferometer or a pair of inter-digital transducers (IDTs) that are deposited directly on top of the host material of PCs. (A thin piezoelectric film beneath the IDTs is necessary if the host material is not piezoelectric.) The satisfactory match between theories and experiments provides a solid foundation for exploring new functionalities and pushing PC concepts into products. Phononic band gaps have been utilized to design PC-based perfect acoustic mirrors and filters [27, 28], acoustic waveguides [29, 30, 31, 25, 32, 33, 34, 35] and couplers [23] thus far.

Negative refraction was first observed in photonic crystals. Due to the analogy between electromagnetic waves and acoustic waves, Yang *et al.* in 2004 demonstrated the negative refraction of acoustic waves upon entering and leaving a three-dimensional PC with flat surfaces [18]. Follow up studies on negative refraction [36, 37, 38, 39, 40] further clarified the forming of the phenomena is caused by the strong anisotropy of the periodic structure at edges of phononic band gaps. Despite having the ability to manipulate the propagation direction of acoustic waves, negative refraction in PCs only exists within narrow frequency ranges and restricted to certain incident angles. To overcome the limitation, Hankansson *et al.* proposed a design of PC lenses via an inverse design method that can focus acoustic waves over a larger bandwidth [41].

1.2.3 Tunablility

With the help of theoretical and experimental techniques, researchers have attempted to control the width and position of phononic band gaps in order to obtain tunable PCs.

In 1999, Caballero *et al.* [42] theoretically and experimentally observed the enlargement of phononic band gaps of two-dimensional square and triangular PCs by introducing an extra scatterer at the center of each unit cell. Two years later, Goffaux and Vigneron theoretically proved that phononic band gaps of a two-dimensional square PC consists of square rods in air could be adjusted by rotating the rods [43]. In 2003, Khelif *et al.* employed the FDTD method to calculate the phononic band gap of two-dimensional square PCs constituted of steel hollow cylinders in water [44]. The position of phononic band gaps is found to be sensitive to the cylinder thickness. In the same year, Hou *et al.* presented a theoretical work that calculates the phononic band gap dependency on the orientation of anisotropic scatterers [45]. Unfortunately, none of the above techniques is feasible for dynamical tuning of the phononic band gaps. In 2007, Yeh proposed a design of two-dimensional PCs composed of isotropic cylinders embedded in an electrorheological material whose elastic properties can be changed by the applied electric fields [46]. Though the tuning range of phononic band gap is small, his proposed structure is considered as the first tunable PC. In 2008, Bertoldi and Boyce presented another design of tunable PC composed of a two-dimensional matrix of circular holes in an elastomeric solid [47]. When stress is applied, the elastomeric material undergoes dramatic physical transformation, which in turn changes the position of phononic band gaps. In 2008, Yang and Chen [48] replaced the steel hollow cylinders in Khelif's design [44] with dielectric elastomer cylindrical actuators whose thickness can be controlled by the applied electric fields. The location and width of the phononic band gaps of their two-dimensional PCs was theoretically proved sensitive to the change in applied external voltage to the hollow actuators. Applications on acoustic switches and tunable acoustic lenses were also computationally demonstrated [48, 49]. Then in 2009, Robillard *et al.* showed that the phononic band gaps of two-dimensional square PCs composed of magnetoelastic constituents can be tuned by application of an external magnetic field [50]. Among the above studies of tunable PCs, phononic band gaps were tuned either by direct physical deformation of the structure or external stimuli (e.g. electric and magnetic fields). The former is impractical for most applications, and the latter often requires very strong stimuli to produce only modest adjustment.

1.3 Overview of the book

This book describes our research efforts on developing acoustic metamaterials which can manipulate the propagation of acoustic waves in a particular manner not found in natural materials and regular PCs. Chapters 2–7 are expanded versions of published, submitted or soon to be submitted technical research manuscripts. A short motivation section is given at the beginning of each chapter to specify the objectives of that particular chapter. A brief literature review is also given to further distinguish the novelty of our work when necessary. Chapter 8 outlines the summary of this book and discusses the prospective directions for future research. Two appendices are included at the end to briefly describe the principle of the PWE and FDTD methods that are heavily used in this book.

1.3.1 Gradient-Index Phononic Crystals

A couple of different GRIN PCs are designed and presented in this book in order to overcome the limitations in manipulating the propagation of acoustic waves by regular PCs. These GRIN PCs can be divided into two groups according to the function of their gradient profiles: linear and hyperbolic secant.

Gradient-Index Phononic Crystals with a Linear Gradient Profile

Chapters 2–3 describe the construction and characteristics of two GRIN PCs with a linear gradient profile. In Chapter 2, we numerically investigate the wide-band collimation phenomena in PC composites, a sequenced series of PCs with the same period but different filling fractions. The methodology and analysis methods described in this chapter serve as the foundation of further PC research that are covered in Chapters 3–6. In Chapter 3, we extend the PC composite concept to form a two-dimensional gradient-index phononic crystal (GRIN PC) whose filling fraction is continuously and linearly modulated. As a result, such a GRIN PC is a direct analogy of gradient media in nature. By controlling the incident angle or operating frequency, a GRIN PC can dynamically adjust the curved trajectory of acoustic wave propagation inside the PC structure to demonstrate the "acoustic mirage" effect on the wavelength scale, without introducing predefined defects as in PC-waveguides.

Gradient-Index Phononic Crystals with a Hyperbolic Secant Gradient Profile

Chapters 4–6 present the formation and characteristics of GRIN PCs with a hyperbolic secant gradient profile. In Chapter 4, we describe the design and characterization of two-dimensional GRIN PCs with a hyperbolic secant profile along the direction transverse to the acoustic wave propagation. The hyperbolic secant gradient distribution is modulated by means of elastic properties of the cylinders. Such a GRIN PC allows subwavelength acoustic focusing over a wide range of working frequencies, making it suitable for applications such as flat acoustic lenses and couplers. In Chapter 5, we report a novel approach to effectively couple acoustic energy into a two-dimensional PC waveguide by an acoustic beamwidth compressor using the concept of hyperbolic-secant GRIN PCs. In Chapter 6, we present the design of a GRIN PC-based acoustic beam aperture modifier to efficiently vary an acoustic beam aperture, i.e. increase or decrease the beam aperture with minimum energy loss and wave form distortion.

1.3.2 Tunable Phononic Crystals

In Chapter 7, we present a theoretical study on the tunability of phononic band gaps in two-dimensional PCs consisting of various anisotropic cylinders in an isotropic host. A two-dimensional PWE method is used to analyze the band structure of the PCs; the anisotropic materials used in this study include cubic, hexagonal, trigonal, and tetragonal crystal systems. By reorienting the anisotropic cylinders, we show that phononic band gaps for bulk acoustic waves propagating in the PC can be opened, modulated, and closed. The methodology presented here enables enhanced control over acoustic metamaterials which have applications in ultrasonic imaging, acoustic therapy, and nondestructive evaluation.

Chapter 2

Wide-Band Collimating by Phononic Crystal Composites

In this chapter, we numerically investigate the wide-band collimation phenomena in PC composites, a sequenced series of PCs with the same period but different filling fractions. The methodology and analysis methods described in this chapter serve as the foundation of further PC research that are covered in the following chapters. Section 2.1 describes the motivation for the study. Section 2.2 explains the formation of self-collimation phenomena in PCs from the equal-frequency contours obtained with a plane wave expansion (PWE) method. Section 2.3 presents the strategy used to overcome the limitations of single PC-based acoustic collimation lenses. In Section 2.4, the finite-difference time-domain (FDTD) method is utilized to simulate the propagation of acoustic waves inside the PC structures. The simulation results show that in comparison to a single PC-based acoustic collimation lens, a PC composite can significantly enlarge the collimation region and realize wide-band acoustic collimation. This work has been reported in *Applied Physics Letters* [17] and featured as the front cover image of the issue.

2.1 Motivation

Flat slabs of PCs have recently been engineered to collimate acoustic waves at designated frequencies where the curvature of an equal-frequency contour (a cross section of a dispersion surface which characterizes all allowed wave vectors corresponding to a same frequency) in wave vector space is nearly zero [6, 38]. Such diffractionless propagation in a PC is described as self-collimating or self-guiding. PC-based acoustic collimation lenses can be made much smaller than traditional collimation lenses with curved surfaces and production of such lenses by micromachining techniques is straightforward and inexpensive. These advantages enable the mass-production of collimation devices at the microscale, where traditional convex and concave lenses are extremely difficult to fabricate.

Despite these advantages, single PC-based acoustic collimation lenses only work within a narrow frequency band where the equal-frequency contours are flat, since the curvature of equal-frequency contour of a PC changes dramatically with frequency. Thus they are not practical for many applications. In this chapter, we present a design concept that is capable of achieving acoustic collimation in a wide frequency range via PC composites.

2.2 Self-Collimation by Single PCs

A beam of acoustic wave is called self-collimated if it does not spread when propagating in a medium. The direction of energy propagation in a medium is given by the group velocity of the wave beam: $v_g = \nabla_k[\omega(\mathbf{k})]$, where ω is the angular frequency and $\mathbf{k}$ is the wave vector. This depicts that acoustic waves propagate along directions normal to the equal-frequency contours, which demonstrate near-zero curvatures at certain frequencies in PC structures. At such frequencies, when incident waves (with varying wave vectors $\mathbf{k}$) end at the flat region of an equal-frequency contour, they are all refracted normal to the flat contour. As a result, the energy fluxes are in the same direction, indicating self-collimation.

There are several methods to calculate the band structure and equal-frequency contours of PCs, including the plane wave expansion (PWE) method [1, 2, 7, 10, 9, 14], the multiple scattering theory (MST) [4, 18, 19, 20, 21, 15, 22], and the finite-different time-domain (FDTD) method [23, 25, 24]. The PWE method is used here due to its strength in solving the vectorial eigenmodes of wave equations for periodic boundary conditions. More detailed information for the PWE method can be found in Appendix A and my previous publication: "Surface and bulk acoustic waves in two-dimensional phononic crystal consisting of materials with general anisotropy" in *Physical Review B* [7]. The accuracy of the PWE method depends on the number of the terms used in the series expansion. In this chapter, 21×21 plane waves are used in the calculation of band structure to reach

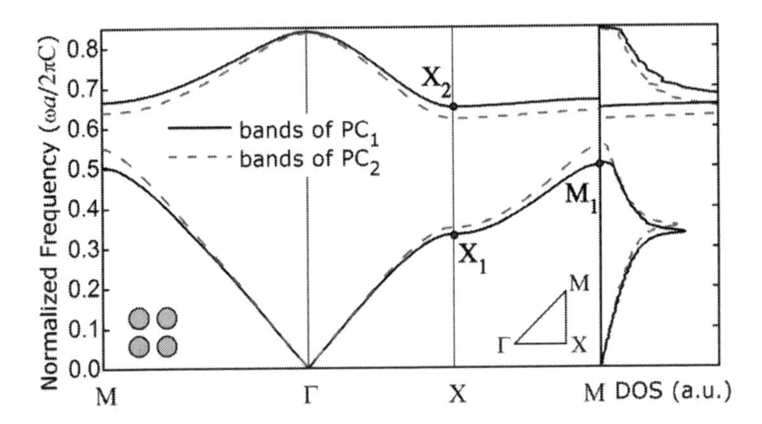

Figure 2.1. Band structure for SV-mode BAW (left) and density of states (DOS, right) of a square-lattice PC composed of steel circular cylinders in water with filling fractions of 0.5454 (PC_1) and 0.4418 (PC_2). The frequency is normalized by $2\pi C/a$, where a is the lattice constant and C is the sound velocity in water.

precise results.

Two single two-dimensional PCs, which we use to investigate self-collimation phenomena, are steel/water PCs consisting two-dimensional square arrays of rigid steel circular cylinders immersed in water. The material parameters are ρ = 7.67 g/cm^3, C_L = 6.01 km/s, C_T = 3.23 km/s for steel, and ρ = 1.0 g/cm^3, C_L = 1.49 km/s for water. ρ is the density and C_L and C_T represent longitudinal and transverse speeds of sound, respectively. The lattice constant a of two single PCs is the same but the filling fraction ($\pi r^2/a^2$, where r is the radius of the cylinder) of the first PC (PC_1) is 0.545 and that of the second PC (PC_2) is 0.442. To simplify the case, we focused our investigation on the shear-vertical (SV) mode BAW where the polarization is parallel to the axes of the cylinders.

Figure 2.1 shows the band structure for the first two SV-mode BAW bands of PC_1 and PC_2. The vertical axis is the normalized frequency defined as $\Omega = \omega a/2\pi C$, where C is the sound velocity in water, and the horizontal axis is the reduced wave vector along the three major orientations of the irreducible Brillouin zone (the smallest enclosed solid figure) in wave-vector space. The differences in filling fractions results in the slope change

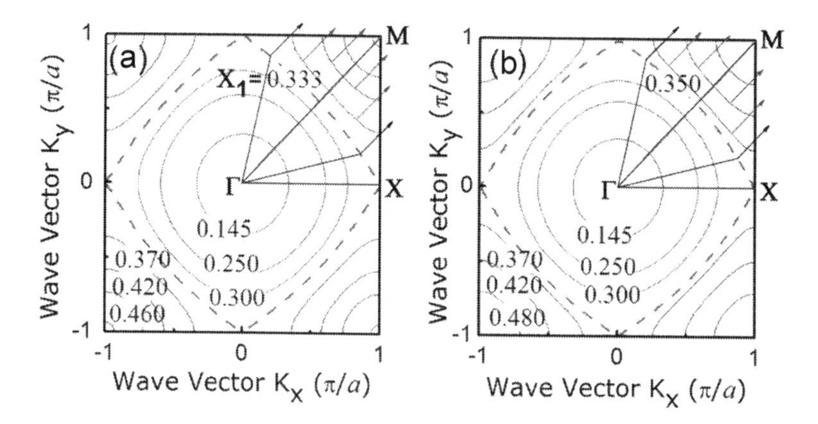

Figure 2.2. (a) and (b) are the equal-frequency contour of PC_1 and PC_2, respectively. The red dotted lines represent the maximum collimation points, and the arrows indicate the boundaries for the collimation region at different frequencies.

in dispersion curves, where a higher filling fraction results in a smaller slope. Both PCs demonstrate a complete band gap ranging from $\Omega = 0.504$ (M_1) to $\Omega = 0.650$ (X_2) for PC_1 and $\Omega = 0.551$ to $\Omega = 0.622$ for PC_2 with relative bandwidths (bandwidth divided by central frequency) of 25.3% and 12.1%, respectively. There is also a partial band gap in the ΓX orientation with normalized frequency Ω extending from 0.333 (X_1) to 0.504 (M_1) for PC_1, and from 0.350 to 0.551 for PC_2.

Figures 2.2(a) and 2.2(b) show the equal-frequency contours for the first band of PC_1 and PC_2, respectively. As the normalized frequency Ω is increased, the curvature of equal-frequency contours of both PCs turns from convex to concave. In this transition process, the flat regions on the equal-frequency contours indicated by the solid arrows which point along the energy-propagating direction (ΓM orientation), enlarge steadily with frequency and then reduce gradually after reaching their maxima ($\Omega = 0.333$ for PC_1 and $\Omega = 0.350$ for PC_2; bold dashed curves). These maxima indicate the frequencies at which the self-collimation angle, the region within which wave vectors undergo self-collimation along the ΓM orientation, are largest. Moreover, the density of states (DOS), number of the compositions of (k_x, k_y) at a certain frequency, in Fig. 2.1 is inversely proportional to the slope of the dispersion curves and hence the curvature of the equal-frequency contours.

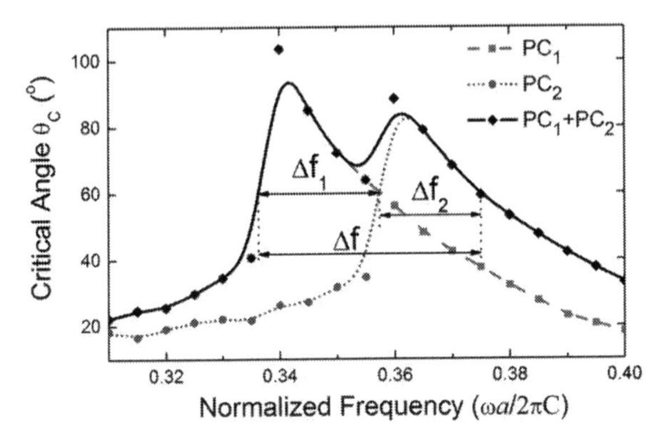

Figure 2.3. Dependence of critical angles at normalized frequency for PC_1, PC_2, and the PC composite (PC_1+PC_2). Δf_1, Δf_2, and Δf are the calculated collimation regions for PC_1, PC_2, and the PC composite, respectively, assuming $\theta_c = 60°$.

Therefore, point X_1 gives the maximum DOS of PC_1, and incident waves with a range of wave numbers can be collimated at point X_1.

2.3 Wide-Band Collimation by PC Composites

Although self-collimation can be achieved with a single PC structure as demonstrated in Figs. 2.1 and 2.2, such collimation system can only work over a narrow frequency range. As the frequency deviates from the maximum self-collimation frequency ($\Omega = 0.333$ for PC_1 and $\Omega = 0.350$ for PC_2), the shape of the equal-frequency contour changes into a circular geometry and only a small portion of the equal-frequency contour depicts a near-zero curvature (Fig. 2.2). Such circular geometries permit only waves within a certain incident angle θ_c to be collimated, thereby limiting the use of single PC-based acoustic collimation lenses. The collimation angle (θ_c)-frequency relations of PC_1 and PC_2 are calculated from the curvature of equal-frequency contours and shown in Fig. 2.3, where the collimation angle is determined as the maximum incident angle within which all refracted waves are less than $\pm 2.5°$ with respect to the ΓM orientation. At $\theta_c = 60°$, the normalized frequency ranges of the collimation regions for PC_1 and PC_2 are calculated to be 0.336–0.357 (Δf_1) and 0.357–0.375 (Δf_2), respectively.

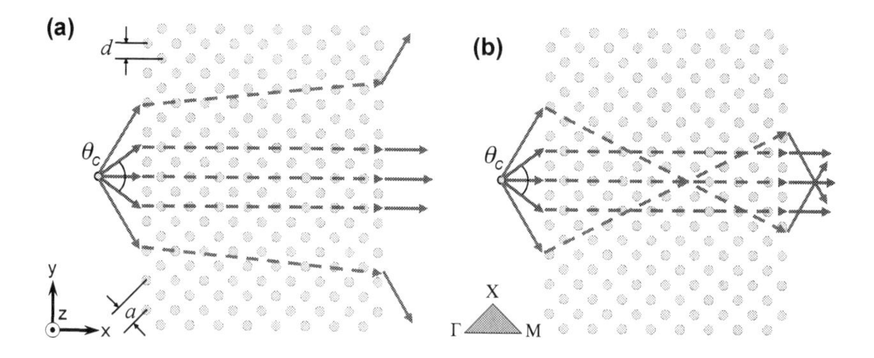

Figure 2.4. The propagation of acoustic waves through a Single PC structure (PC_1 or PC_2) that had homogenous periodicity and homogenous filling fractions at a frequency with (a) convex and (b) concave equal-frequency contour. Arrows indicate the direction of energy transportation.

Figure 2.4(a) illustrates the propagation of acoustic waves that are generated from a point source placed on the left-hand side of a single PC at a frequency with convex equal-frequency contour (smaller than the maximum self-collimation frequency). The solid and dashed arrows indicate the wave propagation directions in water and PC, respectively. While incident waves within the collimation angle θ_c are self-collimated along the ΓM orientation by the PC, incident waves outside θ_c are refracted outwardly due to the convex shape of equal-frequency contour. On the contrary, at frequencies with concave equal-frequency contours (larger than the maximum self-collimation frequency), non-collimated acoustic waves are negative refracted toward the center of the self-collimated beam due to the concave shape of equal-frequency contour, as illustrated in Fig. 2.4(b). Therefore, the deformation of wavefronts after acoustic waves passing through the single PC limits the use of single PC-based optical collimation lenses.

In order to overcome the limitations of single PC-based acoustic collimation lenses, we design a wide-band collimating lens from a series of PCs having the same periods but different filling fractions. Such a PC composite structure can self-collimate acoustic waves in a frequency band which is the summation of the narrow self-collimation bands of each PC that forms the PC composite.

To illustrate the mechanism, a two-dimensional PC composite consisted of two two-

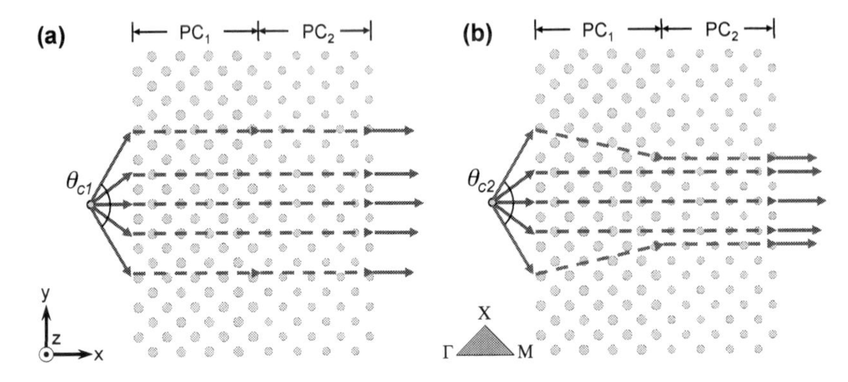

Figure 2.5. (a) Propagation of acoustic waves through a PC composite consisting of two PCs of the same periodicity but different filling fractions when (a) $\theta_{c1} > \theta_{c2}$ and (b) $\theta_{c2} > \theta_{c1}$. Arrows indicate the direction of energy transportation.

dimensional PCs (PC_1 and PC_2 in the previous section) is shown in Fig. 2.5. The cylinders in PC_1 and PC_2 are oriented in a similar fashion, and the first columns of the subsequent PC (PC_2) are positioned exactly at the periodic frame of the previous PC (PC_1). A point source is placed on the left-hand side of the PC composite. For the generated acoustic waves with different frequencies, two cases are involved: (i) when waves pass through the first PC (PC_1) at frequencies near the maximum self-collimation frequency of PC_1 ($\Omega = 0.333$), the incident waves within the collimation angle θ_{c1} of PC_1 are self-collimated along the ΓM orientation and remain unchanged as they propagated in PC_2 [Fig. 2.5(a)]. Since the collimation angle of PC_2 (θ_{c2}) is smaller than θ_{c1} in this frequency range, PC_1 dominates the acoustic collimation behavior and PC_2 retains the self-collimated beam. (ii) When acoustic waves are generated at frequencies near the maximum self-collimation frequency of PC_2 ($\Omega = 0.350$, $\theta_{c2} > \theta_{c1}$ in this case), few of the waves (within collimation angle θ_{c1}) are guided by PC_1 along the ΓM orientation [blue dashed arrows in Fig. 2.5(b)]. The waves outside the collimation angle θ_{c1} are subject to negative refraction as they pass through PC_1 toward the center of the self-collimated beam, due to the concave-like equal-frequency contour (red dashed arrows in PC_1). Not self-collimated by PC_1, the incident waves with angles less than the collimation angle θ_{c2} of PC_2 ($\theta_{c2} > \theta_{c1}$) are subject to self-collimation upon propagating through PC_2 (red dash arrows in PC_2). In this case,

PC_2 is the major contributor to self-collimation, and the function of PC_1 is to confine the out-spreading energy to the center of the self-collimated beam. The PC composite thereby enlarges the self-collimation frequency bands by combining frequency bands of two single-PC structures. As shown in Fig. 2.3, for example, the PC composite significantly enlarges the collimation region by $\Delta f/\Delta f_1 = 185\%$ and $\Delta f/\Delta f_2 = 220\%$ at $\theta_c = 60°$. Such a step-index PC composite (PC_1+PC_2) may serve as the basis of an acoustic collimation lens that focuses waves of a wide frequency range. This idea can be extended to PC composites composed of more than two PCs, and the range of the collimation region can be further improved, with the compensation of higher transmission loss.

2.4 Simulation Results and Discussion

To verify the calculated results based on the PWE method, we employ the FDTD method to simulate the collimation phenomena of a PC composite within the collimation region (Δf=0.336–0.375). In the FDTD program, the equation of motion and the constitutive law are discretized to simulate acoustic wave propagation in linear elastic materials. More information about the FDTD method can be found in Appendix B.

Fig. 2.6(a) is a schematic of the PC composite used in the FDTD simulation. The simulation domain consists of a rectangular area ($30a \times 20a$ normalized units) filled with water and steel circular cylinders distributed periodically with a lattice constant a. PC_1 has 11 layers in the center of the simulation domain and PC_2 has 7 layers at each side of PC_1. The simulation domain is discretized into 24 grids per a, and the radius of steel cylinders in PC_1 and PC_2 is 10 and 9 grids, respectively. The lattice constant a is assumed to be 8 mm, therefore the collimation region (Ω=0.336-0.375) corresponds to a real frequency range of 63.5–68.9 kHz. Other parameters (e.g. material properties) used in the FDTD simulations are the same as those in the PWE calculations. The entire simulation domain is surrounded by perfectly matched layers at the boundaries to avoid reflections. The number of grids and damping coefficient of the perfectly matched layers are 40 and 0.2, respectively. A point source is placed in the center of the structure to generate a SV-mode BAW and the propagation of acoustic waves with time is monitored. A line receiver of length $20a$ is placed at the right edge of the simulation domain, parallel to the y coordinate, to measure the acoustic wave transmittance. The simulation results [Figs. 2.6(b)–(d)] show clear collimation phenomena at three different normalized

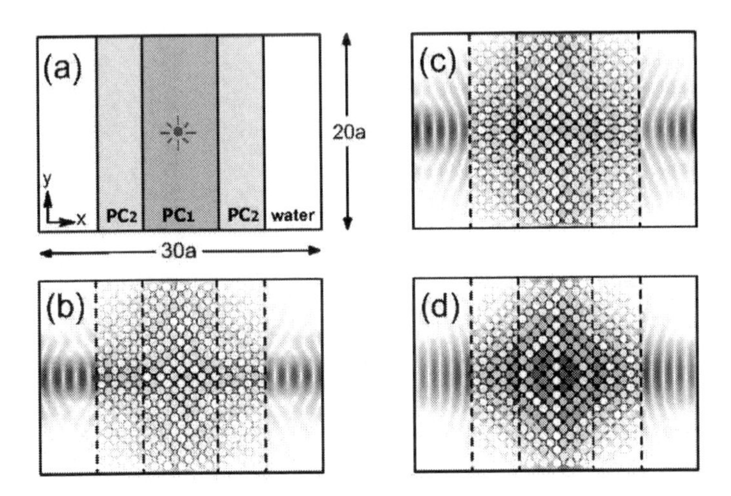

Figure 2.6. (a) Schematic of the PC composite used in the FDTD simulation. Simulated wave propagation inside the PC composite at a normalized frequency of (b) 0.340, (c) 0.355, and (d) 0.370. The light region and the dark one represent weak and strong amplitudes of the displacement field, respectively.

frequencies (0.340, 0.355, and 0.370, corresponding to 63.5 kHz, 66.2 kHz, and 68.9 kHz, respectively) within the collimation region. Quantitative analysis (Fig. 2.7) based on the FDTD simulation results indicate that for all the frequencies in the collimation range (Δf), most acoustic waves are confined in the center region after passing through the PC composite. Thus, the FDTD simulation results coincide with the PWE calculated results and confirm that a PC composite can significantly enlarge the collimation region and cause wide-band acoustic collimation.

2.5 Summary

In this chapter, we have numerically investigated the collimation phenomena of PC composites through both PWE and FDTD methods. The flat bands and the density of states near the collimation point are key factors in determining the collimation characteristics of acoustic waves. An acoustic collimation lens composed of two PCs (steel cylinders in

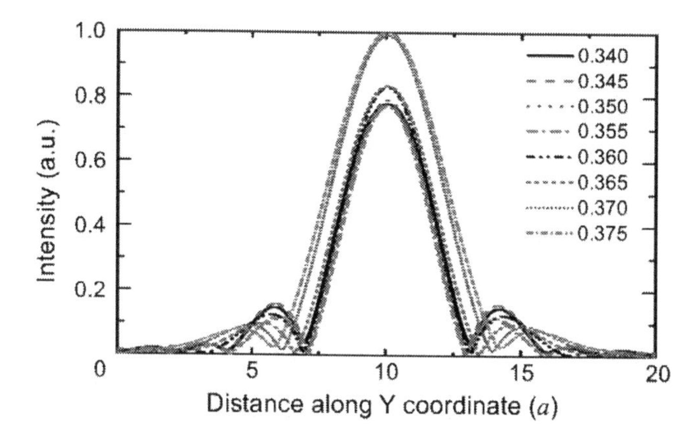

Figure 2.7. Simulated wave intensity at $x = 30a$ along Y coordinate indicates that most acoustic waves at frequencies within Δf are confined in the center region after passing through the PC composite.

water) with different filling fractions is shown to enlarge the collimation region over a normalized frequency range by a factor of 185%–220%, thereby realizing wide-band acoustic collimation. These PC composite-based acoustic collimation lenses will prove useful in applications that require confined acoustic energy flow over long operation distances, such as acoustic imaging, cell sonoporation, and nondestructive evaluation.

Chapter 3

Acoustic Mirage in Gradient-Index Phononic Crystals with a Linear Gradient Profile

In Chapter 2 we have shown that a PC composite has greater flexibility to control the propagation of acoustic waves than single PCs. In this chapter, we extend the PC composite concept to form a two-dimensional gradient-index phononic crystal (GRIN PC) whose filling fraction is continuously and linearly modulated. As a result, such a GRIN PC is a direct analogy of gradient media in nature. By controlling the incident angle or operating frequency, a GRIN PC can dynamically adjust the curved trajectory of acoustic wave propagation inside the PC structure to demonstrate the "acoustic mirage" effect on the wavelength scale, without introducing predefined defects as in PC-waveguides. Section 3.1 introduces the origin of acoustic mirage and describes the motivation for the study. Section 3.2 details the formation of wave-length scale acoustic mirage in a GRIN PC. The propagation trajectories of acoustic waves are predicted from the equal-frequency contours of the GRIN PC calculated by the PWE method. In Section 3.3, the acoustic mirage effect is numerically demonstrated with the FDTD simulation, and the angular and frequency sensitivities of the acoustic mirage effect are studied. Lastly, the design of a tunable acoustic waveguide whose guiding path can be tuned by switching operating frequencies is shown in Section 3.4. The work presented in this chapter has been reported in *Journal of Applied Physics* [51].

3.1 Motivation

PCs are well known for their ability to completely suppress the propagation of acoustic waves within the phononic band gaps. This principle has bee used to construct acoustic waveguides by introducing straight- or bent-line defects to PCs [29, 25, 35]. Due to the fact that acoustic waves are only allowed to propagate in the inclusion-free area, new eigenmodes (called defect modes or guided modes) are generated along with new phononic band gaps within the original phononic band gap of the perfect PC structure. Besides, once defects are defined, the propagation path of acoustic waves is fixed and not adjustable by any means. As a result, the applications of PC-based waveguides are restricted.

Understanding how waves bend under specific condition may reveal a new realm of effective optical/acoustical devices such as on-chip waveguides, filters, and multiplexers. As described by Snell's Law, an atmospheric refractive-index gradient due to temperature variation can bend light waves to project a false image above or below a real object [52]. This phenomenon, known as the optical mirage, is analogous to the acoustic mirage where sound-speed gradients due to temperature, pressure, or salinity can guide acoustic waves traveling in seawater along a bowed trajectory toward lower sound-speed regions [53]. Originally only observed in nature on the kilometer scale due to the low gradient rate of refractive index or sound-speed with respect to distance, recent studies, led by E. Centeno *et al.*, have demonstrated the optic mirage on the wavelength-scale by introducing a sharp refractive-index gradient into the medium [54]. With the potential to transform current acoustic communication systems, the study of the wavelength-scale acoustic mirage effect, an emerging field of acoustics with little prior exposure, holds great promises for realizing acoustic switches and waveguides where a tunable, curved propagation pathway is adapted to control the destination of acoustic waves.

In the next section, we present the design of a two-dimensional GRIN PC with a linear gradient profile, which effectively demonstrate the acoustic mirage effect on the wavelength scale. Using the GRIN PC, the propagating direction of acoustic waves can be continuously bent along an arc-shaped trajectory by gradually tuning the filling fraction of phononic crystals.

3.2 Beam Bending in Two-Dimensional GRIN PCs

It is known that when an acoustic velocity mismatch exists between a homogeneous slab and its surrounding medium, refraction occurs at the entrance and exit interfaces of the slab. The amount of beam bending is dependent upon the velocity difference, and it can be investigated through the study of equal-frequency contours in the wave-vector space. In contrast to the circular equal-frequency contours of a homogeneous, isotropic medium, the shape of the equal-frequency contours of a PC can be highly anisotropic and frequency dependent when the acoustic wavelength is on the scale of the structure's periodicity. As a result, large-angle beam bending can be realized in PCs set at designated frequencies. By calculating the effective acoustic velocity for the propagation wave mode of a PC from its equal-frequency contours, one can readily predict the direction of refraction using Snell's law if anisotropy of the PC is negligible [54]. However, once the direction of refraction is determined, the refracted acoustic waves propagate along a straight line through the PC structure until entering a medium of different acoustic velocity. Thus, existing perfect-periodic PCs cannot redirect acoustic waves continuously to exhibit the acoustic mirage effect. In order to achieve a bent path for acoustic wave propagation, the direction of the group velocity must be successively modified during the propagation.

An acoustic-velocity gradient can be obtained in a GRIN PC by locally modulating the constitutive parameters (e.g., geometry, filling fraction, material properties) of the structure. Since the band structure of a PC is highly sensitive to the variation of its constitutive parameters, the shape of the equal-frequency contours of a PC can be deformed by changing the inclusion radius, which in turn changes the acoustic velocity and refraction angle of the acoustic beam. By extending the PC composite concept introduced in the previous chapter, here we present the design of a linear-index GRIN PC whose filling fraction along the wave propagation direction is continuously modified.

As illustrated in Fig. 3.1, a two-dimensional GRIN PC with an one-dimensional gradient is a discretized medium that can be thought of as a composite of multiple single-layer PCs of different filling fractions. When an acoustic beam propagates through the two-dimensional GRIN PC, it encounters redirection at every virtual interface between layers, resulting in consecutive reorientations of the acoustic beam inside the structure. Thus, by gradually modulating the filling fraction of a GRIN PC, one may create an arc-shaped trajectory for acoustic wave propagation. The results presented in this work are restricted to

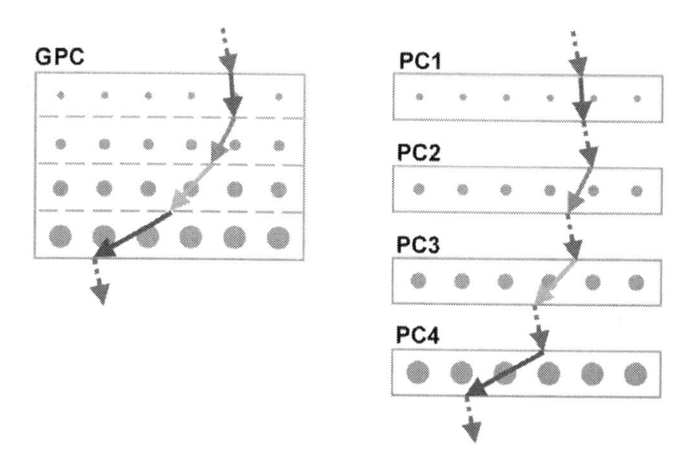

Figure 3.1. A smooth redirection of acoustic wave propagation can be achieved by a GRIN PC where each layer is considered an independent PC of different filling fraction.

an inclusion-radius gradient, but the concept can be employed for the inclusion-property or lattice-spacing gradients as well.

We first study the effect of the filling fraction variation on the deformation of equal-frequency contours of two-dimensional PCs. Figure 3.2 shows the calculated band structure for the SV-mode BAW of three square-lattice PCs (PC_1, PC_2, and PC_3) with perfect periodicity, each consisting of steel cylinders embedded in epoxy. Their filling fractions are 1.2%, 3.4% and 6.7%, respectively. The band structure is obtained using the PWE method, assuming the two-dimensional PCs are infinitely periodic in the x and y directions while the SV-mode waves are polarized along the z direction (the cylinder axis). The material properties for steel are the same as used in Chapter 2, and those for epoxy are $\rho = 1.14$ g/cm^3, $C_L = 2.55$ km/s, $C_T = 1.14$ km/s. From the band structure shown in Fig. 3.2 we observe that the differences in filling fractions results in the slope change in dispersion curves, where a higher filling fraction results in a smaller slope. This trend matches our observation from Fig. 2.1, though the background is water in Chapter 2 and epoxy here.

Figure 3.3(a) displays the three-dimensional dispersion curves of the second band of PC_1 and PC_3. The light (red) and dark (black) solid contours represent the equal-frequency contours of the PCs at a reduced frequency $\Omega = 0.79$. The bell-shape of the

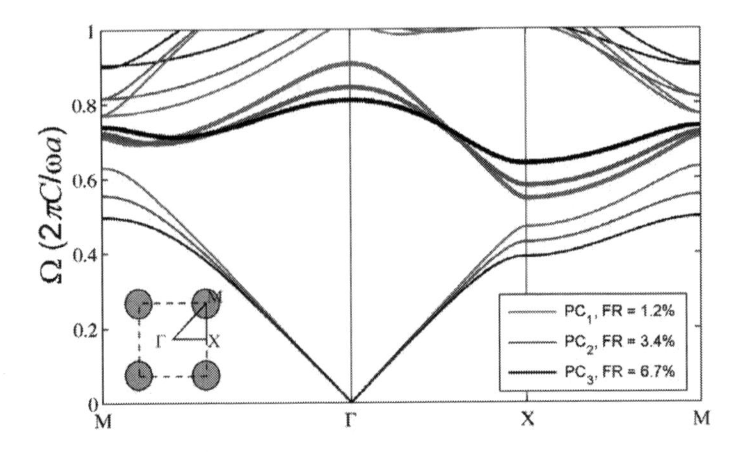

Figure 3.2. Band structure for the SV-BAW mode of three PCs with filling fractions of 1.2% (PC_1), 3.4% (PC_2), and 6.7% (PC_3).

dispersion curves of both PCs implies that the group velocity points inwards, indicating negative refraction. As the filling fraction increases from 1.2% to 6.7%, the peak of the dispersion curve drops and the size of the equal-frequency contour decreases. As a result, the group velocity within the PC of a higher filling fraction is greater than that within the PC of a lower filling fraction.

Figure 3.3(b) shows equal-frequency contours for epoxy and three square-lattice steel/epoxy PCs with filling fractions of 1.2%, 3.4%, and 6.7% at a reduced frequency of 0.79. The dotted arrow represents the incident acoustic beam with an angle of 10° with respect to k_y. By considering the conservation of the transverse momentum, we obtain a construction line [thin vertical line in Fig. 3.3(b)] that intersects with all equal-frequency contours. The refraction direction of the acoustic beam in each PC points in the direction of the inward arrow, which is normal to the equal-frequency contour. It is shown that small perturbations in filling fraction produce a large variation in the refraction angle. Hence, by steadily increasing the filling fraction of a PC, one can alter the shape of the second-band equal-frequency contour from concave to convex, causing the acoustic beam to gradually redirect toward the center of the Brillouin zone (Γ-point).

To find an equation for the approximated curved trajectory of optical/acoustical mi-

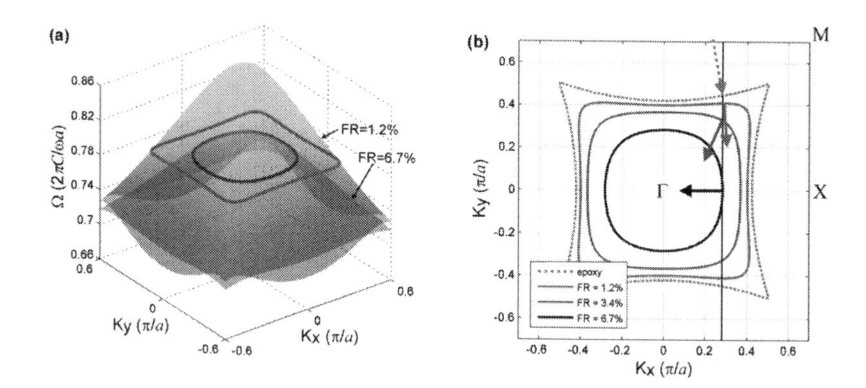

Figure 3.3. (a) Three-dimensional dispersion curves of the second band for the SV-BAW mode of PC_1 and PC_3. The red and black curves represent the equal-frequency contours at a reduced frequency of 0.79. (b) equal-frequency contours at a reduced frequency of 0.79, for epoxy and PCs with three different filling fractions. The dotted arrow represents the incident direction. The solid line represents the construction line plotted for an incident angle of 10°. The solid arrows represent the direction of each group velocity.

rage effects in nature, the Eikonal equation is used to describe the wave propagation in a piecewise isotropic medium [52]

$$\nabla n(\mathbf{u}) = \frac{d}{ds}\left[n(\mathbf{u})\frac{d\mathbf{u}}{ds}\right], \tag{3.1}$$

where $\mathbf{u}$ is the ray vector, n is the refractive index, and s is an elementary path length. Equation 3.1 is useful for describing the trajectory of the acoustic mirage effect in a GRIN PC that operates at the frequencies where the equal-frequency contours are nearly circular (e.g., at lower frequency region of the first-band for the SV-BAW mode). However, the change in the refraction angle in the first band is less dramatic with change in filling fractions than that in the second band. To obtain sharp continuously bending of an acoustic beam, the second band dispersion curves are chosen to form the GRIN PC. Unfortunately, a GRIN PC would be very anisotropic when operates in the second band of SV-BAW that involves negative refraction [Fig. 3.3(b)]. As a result, the Eiknoal equation cannot accurately describe the acoustic mirage effect in this GRIN PC, though it can still serve as quick design guidance. However, it can be predicted from the shape of equal-frequency contour at every layer. As a GRIN PC is not strictly periodic, the local dispersive properties of the GRIN PC cannot be precisely obtained by the ordinary PWE method. However, for a weak gradual variation of the constitutive parameters (e.g., filling fraction in this case), the local dispersive properties of each row of a GRIN PC can be approximated by perfect periodic PCs. Therefore, the curved trajectory of an acoustic beam propagating in a GRIN PC can be predicted by analyzing the local equal-frequency contours of the rows consecutively crossed by the acoustic beam. Numerical simulations in the following paragraphs examine this assumption.

3.3 Acoustic Mirage in GRIN PCs

To numerically demonstrate this superbending phenomenon, we design a two-dimensional GRIN PC that is composed of 25×12 layers of steel cylinders arranged in a square lattice and embedded in epoxy (Fig. 3.4). The radius of each row of cylinders increases linearly with the row number (along the y direction) per the following relation: $r_y = \sigma(y+1)a$, where $\sigma = 2.08\%$. The material properties are the same as those in the PWE calculations. We anticipate that the propagation of the acoustic beam inside the GRIN PC would continuously bend due to the filling fraction gradient. To investigate the propagation of acoustic waves in the GRIN PC, we simulate the SV-mode BAW propagation by the FDTD method. The simulation domain is discretized into 48 grids per a. The lattice constant a is assumed to be 8 mm, therefore $\Omega = 0.79$ corresponds to a real frequency of 112.4 kHz. A tilted line source with a width of $4a$ is placed in the epoxy region (centered at $y = -5a$) and is defined by setting an initial value of body force in the equation of motion along the z axis (parallel to the cylinder axis) to generate a SV-mode BAW. The entire domain is surrounded by 40-grid-thick perfectly matched layers at the boundaries with a damping coefficient of 0.2 to avoid numerical reflections in the simulation domain, resulting in total absorbance at the boundary. To be consistent with the PWE calculations, the thermal effects on material properties are not considered in the FDTD simulations. Other parameters used in the FDTD simulations are the same as those in the PWE calculations.

As shown in Fig. 3.4, the GRIN PC is exposed to an acoustic beam with an incident angle of 10° at $\Omega = 0.79$ ($\lambda = a/\Omega = 1.26a$). The light and dark regions represent strong and weak amplitudes of the displacement field, respectively. The acoustic beam bends

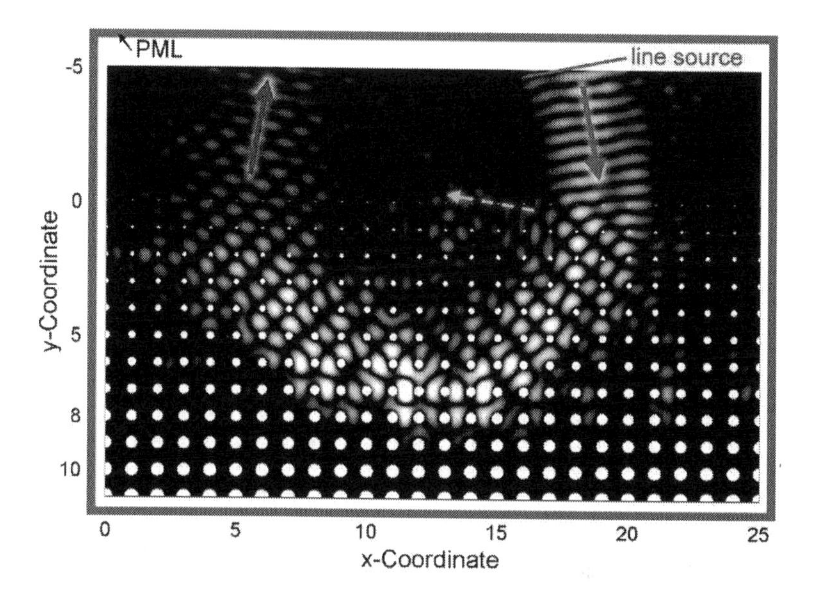

Figure 3.4. "Acoustic mirage" inside a GRIN PC illuminated by a SV-mode acoustic beam with a width of $W = 4a$, an incident angle of 10°, and an operating frequency of $\Omega = 0.79$. The x and y coordinates are in unit of a. The light and dark regions correspond to the strong and weak amplitudes of the displacement field, respectively.

slightly toward the direction of the gradient upon entering the GRIN PC, as predicted from the equal-frequency contours calculated by the PWE method [red arrow in Fig. 3.3(b)]. The beam then steadily refracts toward the negative x direction, redirecting back to the epoxy/GRIN PC interface with a lateral shift of $13.75a$ (equivalent to 10.9λ). The penetrating depth of the acoustic beam at the beam center, measured to be $7a$ (equivalent to 5.5λ), corresponds to a perfect periodic PC with a filling fraction of 6.7% [the dark (black) contour in Fig. 3.3(a)].

While traveling within the GRIN PC structure, the beam exhibits little spread with regard to distance; however, as the beam approaches the epoxy/GRIN PC interface, the width of the output beam widens marginally due to multiple reflections near the interface. A reflection beam propagating toward the negative x direction near the structure surface, denoted by the dashed arrow in Fig. 3.4, is caused by the periodicity of the epoxy/GRIN

PC interface. This reflection beam interferes with the outgoing beam and results in the doted shape of the output field. The transmission of displacement amplitude measured at the output is 60%, matching well with the transmission coefficient calculated from the effective acoustic impedance: $Z_{eff} = Z_{steel}f + Z_{epoxy}(1-f)$, where Z is the acoustic impedance defined as the product of density and acoustic velocity of a material, and f is the filling fraction.

The arc-shaped trajectory in Fig. 3.4 agrees perfectly with the directions of refraction obtained via the PWE method [Fig. 3.3(b)], implying that the trajectory inside a GRIN PC can be predicted by analyzing the local dispersion curves of the structure. Moreover, the results show that the GRIN PC can redirect acoustic waves within conducting bands no matter where the incident beam hits the surface. This characteristic presents a significant advantage over existing PC waveguides, which rely on predefined line defects to guide acoustic waves within band gaps. Here we present the acoustic mirage effect for SV-mode BAW only, but the concept can be directly applied to achieve acoustic mirage for other acoustic wave modes.

3.4 GRIN PC-based Tunable Acoustic Waveguides

Given that equal-frequency contours are anisotropic and frequency-dependent, the propagation trajectory within a GRIN PC would be sensitive to incident angle and operating frequency. Figure 3.5(a) and 3.5(a) show how penetration depth and mirage distance (i.e., the distance between input and output beams) are dependent on incident angle and frequency, respectively. In Fig. 3.5(a), the angular dependence of the penetration depth and mirage distance are both linear at $\Omega = 0.79$, yielding a sensitivity of $0.0625a$ and $0.125a$ per degree, respectively. In contrast, the penetration depth and mirage distance have stronger dependency on the operating frequency at an incident angle of 10°, as shown in Fig. 3.5(b). These results demonstrate that by taking advantage of the angular or frequency sensitivity to select different trajectories within the GRIN PC, tunable acoustic waveguiding can be achieved.

To confirm the tunability of GRIN PC-based waveguiding, the propagation of acoustic waves in a modified GRIN PC with 25×17 layers is simulated. As shown in Fig. 3.6(a), the modified GRIN PC is symmetric with respect to the neutral axis (the horizontal dashed line along $y = 8a$) while the upper part is identical to the top nine rows (y=0–$8a$)

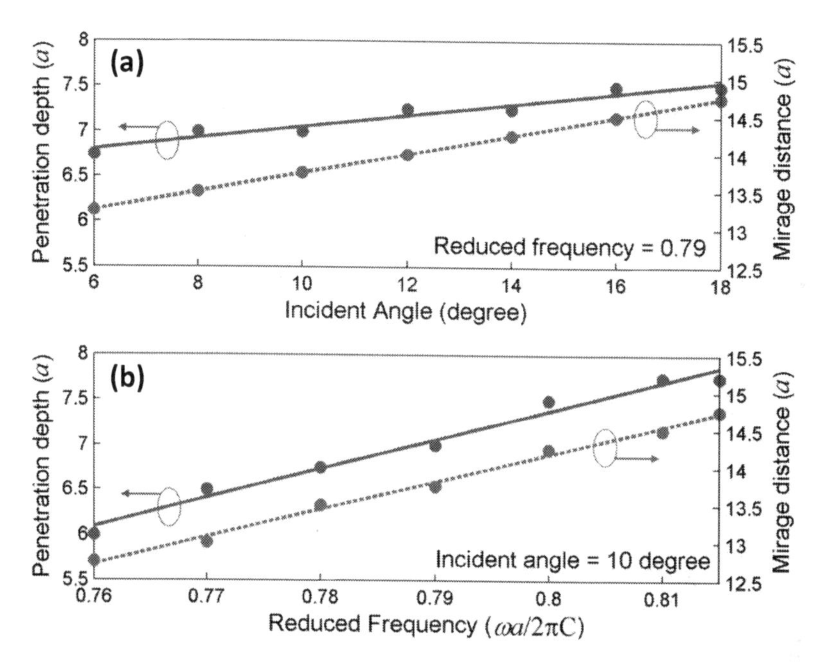

Figure 3.5. (a) Angular and (b) frequency sensitivities in a function of penetration depth and mirage distance in a GRIN PC. The solid red, solid blue, and dashed arrows denote input, output, and reflected beams, respectively.

of the structure used in Fig. 3.4. For an acoustic beam operating at $\Omega = 0.76$ and an incident angle of 10°, the trajectory is concave upward [Fig. 3.6(a)] and similar to that in Fig. 3.4. As the operating frequency is tuned to $\Omega = 0.815$, however, the incident beam is negatively reflected upon crossing the epoxy/GRIN PC interface. The penetration depth increases and the acoustic beam is guided through the structure, as shown in Fig. 3.6(b). The dashed arrow denotes the reflected beam from the epoxy/GRIN PC interface that destructively interfered with the incident beam and produced a gap between two beams. The results confirm that GRIN PC-based waveguides can dynamically control the destination of guided acoustic waves without predefined line defects; this characteristic makes GRIN PC-based waveguides useful in applications such as acoustic switching, filtering and biosensing. Additionally, the transmission of displacement field can be improved when

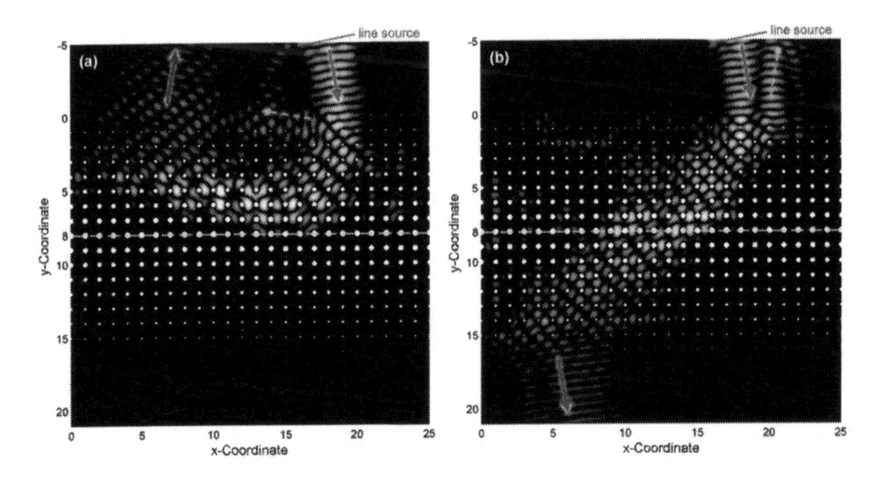

Figure 3.6. Simulated waveguiding inside a modified GRIN PC exposed to a SV-mode acoustic beam of width $W = 4a$, incident angle of $10°$ and reduced frequencies of (a) 0.76, or (b) 0.815. The solid red, solid blue, and dashed arrows denote input, output, and reflected beams, respectively.

the structure is optimized for specific applications.

3.5 Summary

In this chapter, we have computationally shown that the propagation direction of an acoustic wave beam within a GRIN PC can be continuously tuned by modifying the constitutive parameters of the structure. Our parametric analysis of the deformation of the dispersion curves along with FDTD studies have revealed that the acoustic mirage phenomenon can be predicted from local equal-frequency contours. We have found that the propagating trajectory of acoustic waves is sensitive to both incident angle and operating frequency. Based on these findings, we have designed an acoustic waveguide that cannot only operate without line defects, but can also be actively tuned. The methodology described in this chapter can be implemented in acoustic devices such as on-chip waveguides, biosensors, filters, and multiplexers.

Chapter 4

Subwavelength Acoustic focusing by Gradient-Index Phononic Crystals with a Hyperbolic Secant Gradient Profile

In this chapter, we describe the design and characterization of two-dimensional GRIN PCs with a hyperbolic secant profile along the direction transverse to the acoustic wave propagation. The hyperbolic secant gradient distribution is modulated by means of elastic properties of the cylinders. Such a GRIN PC allows subwavelength acoustic focusing over a wide range of working frequencies, making it suitable for applications such as flat acoustic lenses and couplers. This chapter is organized as following. Section 4.1 describes the motivation for the study. Section 4.2 details the general design principle of GRIN PCs with a hyperbolic secant gradient profile. In Section 4.3, the subwavelength acoustic focusing is numerically demonstrated with the FDTD simulation over a wide frequency range. A summary and extension readings are given in Section 4.4 and Section 4.5, respectively. The work presented in this chapter has been reported in *Physical Review B* [55].

4.1 Motivation

Acoustic waves, like electromagnetic waves, are tricky to manipulate on small scales. Scientists have been attempted to build acoustic and optical lenses that can overcome the diffraction limit for subwavelength focusing. Utilizing the unique negative refractions exhibited by PC slabs, subwavelength acoustic focusing has been recently achieved by [18]. This discovery has paved the way for PC-based acoustic lenses since traditional acoustic lenses with curved surfaces are difficult to fabricate, prone to misalignment, and most importantly are confined by the diffraction limit.

Despite the promise of PC-based flat lenses, negative refraction in PCs has been observed only within narrow frequency bands and for small incident angles. As such, PC lenses that focus acoustic waves by means of negative refraction feature restricted ranges of operation, large focal zones, and low gain. Moreover, existing PC lenses are bulky since they operate within the first partial band gap or within higher-order dispersion bands. The aforementioned limitations call for the development of a compact flat acoustic lens that operates over wide ranges. Hakansson *et al.* recently presented a design of PC lenses via inverse design method that does not involve negative refraction [41]. However, they are still limited to narrow working band and bulky size. In this chapter, we demonstrate a wide-band flat PC lens by means of two-dimensional GRIN PCs with a hyperbolic secant refractive-index profile.

4.2 GRIN PCs with a Hyperbolic Secant Gradient Profile

In designing a GRIN PC structure we select a refractive index profile that enables acoustic wave focusing. The simplest profile is of parabolic form, and it is widely studied and used in optical devices. We choose a refractive index profile in the form of a hyperbolic secant (of which a parabolic profile can be considered as the first-order Taylor series approximation) [56]. One can analytically solve the equations that govern wave propagation within the hyperbolic secant gradient media; this solution is not an approximation, causing minimum aberration [52]. The refractive index profile of a two-dimensional, continuous GRIN medium along the transverse direction (y axis) is defined as

$$n(y) = n_0 sech(\alpha y), \tag{4.1}$$

where n_0 is the refractive index along the center axis (x axis) and α is the gradient coefficient.

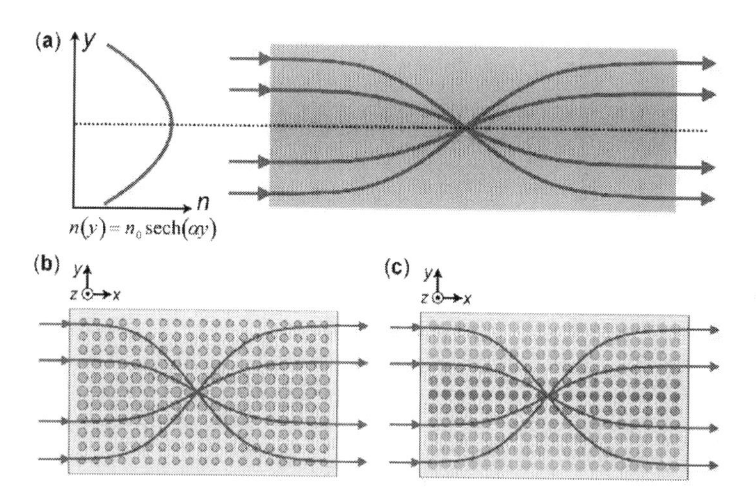

Figure 4.1. (a) Principle of a gradient-index medium. A hyperbolic secant refractive index profile along the direction transverse to propagation (y axis) enables redirection of incident beams (red arrows) inside the medium. A GRIN PC can be realized by (b) adjusting radii of cylinders or (c) changing elastic properties of cylinders along the transverse direction to achieve a hyperbolic secant refractive index profile.

Within the GRIN medium, the paraxial incident beams bend gradually toward the center axis where refractive index is highest per Snell's law (Snell's law for acoustics relates the angles of incidence and refraction for a wave between two media with different material velocities) and converge at a focal spot. Beyond that spot, the focused beams are redirected parallel to the direction of propagation, as shown in Figure 4.1(a). A flat GRIN medium can thereby serve as a flat lens for focusing or for collimation. The beam trajectory within a GRIN medium can be analytically derived from the hyperbolic secant refractive index profile [52]:

$$y(x) = \frac{1}{\alpha} sinh^{-1}[u_0 H_f(x) + \dot{u}_0 H_a(x)], \tag{4.2}$$

where u_0 is the hyperbolic space at $x = 0$ such that $u_0 = sinh(\alpha y_0)$; $\dot{u}_0$ is the derivative of u_0 with respect to x; and $H_f(x)$ and $H_a(x)$ are the positions of axial and field rays

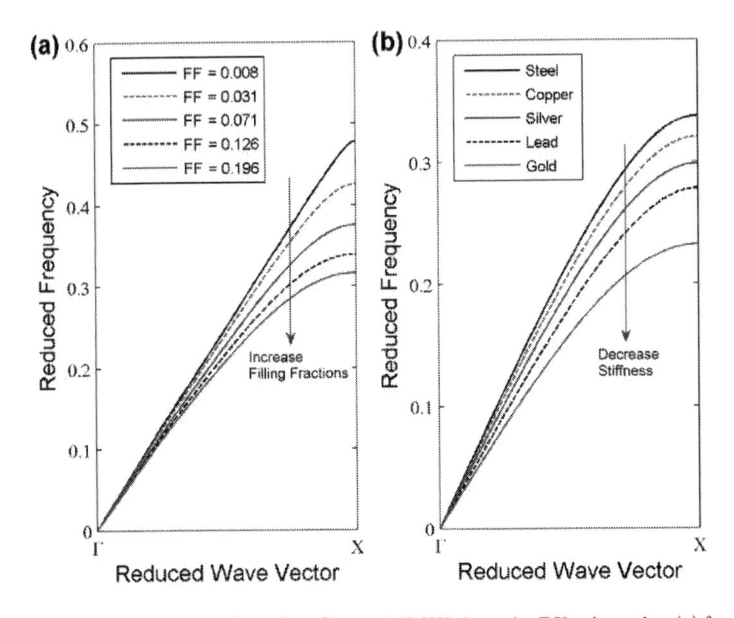

Figure 4.2. Band structure of the first SV-mode BAW along the ΓX orientation (a) for various filling fractions (FFs) of a steel/epoxy PC, and (b) for PCs consist of different solid cylinders in epoxy with a filling fraction of 0.126.

written as

$$H_a(x) = sin(\alpha x)/\alpha, H_f(x) = cos(\alpha x). \tag{4.3}$$

The sinusoidal path through the GRIN medium described by Eq. 4.2 features a focal length (the distance to focus paraxial incident beams within the GRIN medium) of

$$f = \pi/2\alpha. \tag{4.4}$$

The focal length is independent of the incident position along the y direction (y_0). One may thus characterize focusing within a GRIN medium by the gradient coefficient α.

A two-dimensional GRIN PC can be considered as a transversely discretized medium where each row is independently homogeneous and isotropic. One can physically realize this discretized medium by adjusting the refractive index in each row to fit a hyperbolic secant profile. For example, the adjustments can be realized by changing the radiuses or elastic properties of the cylinders in each row, as illustrated in Figs. 4.1(b) and 4.1(c), respectively. Though PCs are anisotropic and frequency-dispersive, the effective refractive index for a certain wave mode of a PC that is driven at a specific frequency can be approximated as the average of the refractive indices along its principal orientations [54, 19]. For small anisotropic ratios ($r_a = v_{\Gamma X}/v_{\Gamma M}$) where equal-frequency contours are nearly circular, the effective refractive index for the considered wave mode of a square-lattice PC is given by

$$n_{eff} = \frac{n_{\Gamma X} + n_{\Gamma M}}{2}. \tag{4.5}$$

Here $n_{\Gamma X}$ and $n_{\Gamma M}$ are refractive indices along the ΓX and ΓM orientations, respectively:

$$n_{\Gamma X} = \frac{C}{v_{\Gamma X}} = \frac{C}{d\omega/dk_{\Gamma X}}, \tag{4.6}$$

$$n_{\Gamma M} = \frac{C}{v_{\Gamma M}} = \frac{C}{d\omega/dk_{\Gamma M}}, \tag{4.7}$$

where v is the group velocity of the considered wave mode and C is the sound velocity of the propagation mode in background material. We calculate the band structure of PCs by the PWE method to determine the refractive indices. Fig. 4.2 shows the band structure of the first SV-mode BAW along the ΓX orientation for various filling fractions of a steel/epoxy PC [Fig. 4.2(a)], as well as for PCs of different solid cylinders in epoxy with a filling fraction of 0.126 [Fig. 4.2(b)]. The group velocity of SV-mode BAW of a PC decreases (and the refractive index increased) when the filling fraction is increased or the cylinder stiffness is decreased. A two-dimensional GRIN PC can thus be formed by carefully adjusting cylinder radii or choosing appropriate materials in different rows to match a hyperbolic secant refractive index profile.

In Chapters 2 and 3, we have demonstrated the design of two linear-gradient GRIN PCs whose gradient profile is obtained by locally adjusting the radius of cylinders. Here we design a two-dimensional GRIN PC with a hyperbolic secant gradient profile that is composed of different cylinder materials along different rows. We form a GRIN PC consists of 19 rows of solid cylinders ($y = [-9a, +9a]$) arranged in a square lattice and embedded in epoxy, with a filling fraction of 0.126. The elastic properties of the cylinders and the epoxy are listed in Table 4.1. The first-band dispersion curves of each row for the first SV-mode BAW are calculated by the PWE method and plotted in Fig. 4.3.

Material	y coordinate	Density (g/cm^3)	V_l (km/s)	V_s (km/s)	N_{eff} ($\Omega = 0.1$)
Cadmium	0	8.6	2.8	1.5	1.186
Molybdenum	±1	10.1	6.29	3.35	1.184
Titanium carbide	±2	10.15	6.7	4.0	1.178
Copper	±3	8.96	4.66	2.24	1.168
Niobium	±4	8.57	4.92	2.1	1.160
Brass	±5	8.1	4.7	2.1	1.144
Cast Iron	±6	7.7	5.9	2.4	1.123
Zinc	±7	7.1	4.2	2.41	1.102
Chromium	±8	7.0	6.65	4.03	1.073
Zinc oxide	±9	5.68	6.33	2.95	1.042
Epoxy		1.14	2.55	1.14	1.000

Table 4.1. Elastic properties of the materials used in the GRIN PC in Sec. 4.2. The effective refractive indices are obtained at reduced frequency $\Omega = 0.10$ with the PWE method, per Eq. 4.5.

The lowest initial frequency of the first partial band gap among all rows is $\Omega = 0.31$, suggesting that the GRIN PC can operate over a wide band from $\Omega = 0.0$–0.31. For frequencies below $\Omega = 0.25$, the dispersion curves are very linear along both ΓX and ΓM orientations, and the equal-frequency contours of all rows are nearly perfect circles. Fig. 4.4 shows the equal-frequency contours of all row of the GRIN PC at $\Omega = 0.1$. The maximum anisotropic ratio ($r_a = v_{\Gamma X}/v_{\Gamma M}$) is less than 1.025 for frequencies below $\Omega = 0.25$, implying that each row can be approximated as an isotropic medium.

The effective refractive indices obtained at reduced frequency $\Omega = 0.10$, per Eq. 4.5, are listed in Table 4.1 and plotted in Fig. 4.5 (red dots) along with a best-fit to the hyperbolic secant profile (solid black line). The effective refractive index of each row is a function of frequency because PCs disperse with frequency. Fitted profiles for different reduced frequencies (Fig. 4.5) show that the gradient coefficient of the GRIN PC is dependent upon reduced frequency. Per Eq. 4.4, one can calculate the theoretical focal lengths at different reduced frequencies. In the next section we show a comparison analysis of these theoretically calculated focal lengths and those from the FDTD-simulated results.

4.3 Subwavelength Acoustic focusing by GRIN PCs

To investigate the capability of a GRIN PC in manipulating acoustic waves for subwavelength focusing, we simulate the propagation of SV-mode BAWs by the FDTD method.

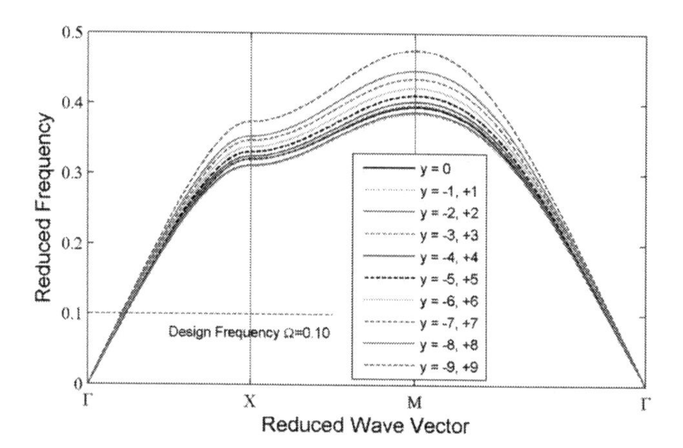

Figure 4.3. The first-band dispersion curves of each row of the GRIN PC for the first SV-mode BAW. The material properties are listed in Table 4.1.

The simulation domain used for FDTD is a two-dimensional GRIN PC that comprises 19 rows in the y direction and 50 layers in the x direction. The material properties of cylinders and the background are listed in Table 4.1, the same as those used in the PWE calculation. An infinitely long line source is placed at $x = 0$ to generate a planar SV-mode BAW. The lattice spacing a is set as 8 mm and each lattice is discretized into 20 uniform spatial grids. The entire domain is surrounded by 40-layer-thick perfectly matched layers at the boundaries to avoid numerical reflections in the simulation domain. The other parameters used in the FDTD simulations are the same as those in the PWE calculations of Sec. 4.2.

Fig. 4.6(a) shows the simulated acoustic wave propagation at reduced frequency $\Omega = 0.05$ along with the beam trajectory calculated from analytic expression in Sec. 4.2. The amplitudes of displacement fields are shown in decibels (dB). The color distribution is normalized per the maximum value of the displacement. The figure shows the redirection of plane waves as they propagate within the GRIN PC and focus at $x = 30a$. This redirection matches with the theoretical value given by $f = \pi/(2 \cdot 0.0522/a) = 30.09a$. The acoustic beams incident upon various y positions has different propagation routes toward the focal spot, but their propagation in the x direction is constant because the

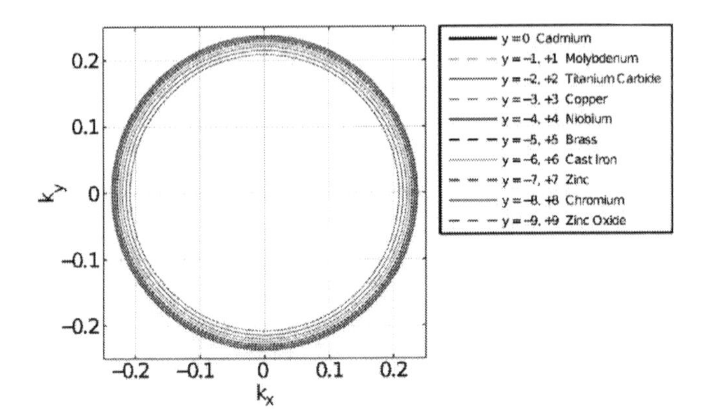

Figure 4.4. Equal-frequency contours of all row of the GRIN PC at the design frequency $\Omega = 0.1$.

velocity of sound is inversely proportional to effective refractive index. As such, there is a token degree of longitudinal aberration within a GRIN PC. The focal spot has a -3 dB width (in the y direction) of $7a$, equivalent to 35% of the wavelength of SV wave in epoxy at $\Omega = 0.05$ ($\lambda = a/\Omega = 20a$), indicating minor transverse aberration. Fig. 4.6(b) shows the propagation of acoustic waves at $\Omega = 0.20$. The structure gradually deforms the planar wavefronts into circular wavefronts, focusing the acoustic beams at $x = 21$ and collimating waves that would have otherwise diverged beyond that point. The -3 dB width of the focal spot is 55% of the wavelength. Figs. 4.6(a) and 4.6(b) illustrate effective acoustic focusing over a wide range of $\Omega = 0.05$–0.20 ($\Delta\Omega = 0.15$), several times wider than the working band of a negative-refraction-based PC ($\Delta\Omega \leq 0.04$) [18, 36, 37, 39], and the focal spot is smaller than wavelength.

The focal length of the GRIN PC versus the operation frequency is presented in Fig. 4.7. The solid and dotted lines represent the focal length calculated from gradient coefficients and obtained from FDTD simulations, respectively. The analytic solutions are close to the simulation results at reduced frequencies below $\Omega = 0.10$. As the reduced frequency increased, the analytic results deviated from the simulated results. This deviation is most likely caused by the fact that at higher frequencies, the wavelength is comparable to lattice spacing. At small wavelengths one cannot assume that the gradient of the GRIN PC is smooth. The fact that equal-frequency contours are less circular at higher

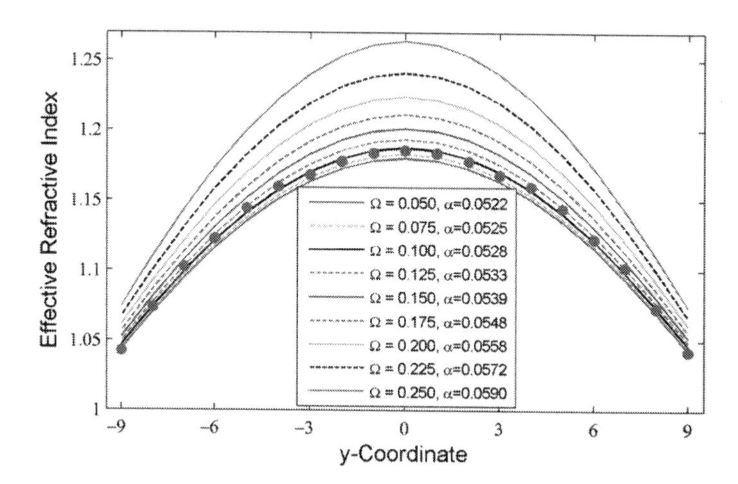

Figure 4.5. Fitted hyperbolic secant profiles for different reduced frequencies. The red dots represent the effective refractive indices at a frequency of $\Omega = 0.10$. The gradient coefficient is found to be sensitive to frequency.

frequencies also contributes to the deviation. As a result, the analytic beam trajectory is only valid when the SV-mode wavelength in epoxy is at least ten times larger than the lattice spacing of the GRIN PC but can still serve as guidance.

The simulated acoustic propagation in GRIN PC at $\Omega = 0.10$ is shown in Fig. 4.8(a), as well as the beam trajectory calculated from Eq. 4.2. The displacement across the y section of the focal spot at $x = 29.5a$ is compared with that of an unfocused acoustic beam centered at $x = 2.5a$, as shown in Fig. 4.8(b). We also plot in this figure the displacements of acoustic waves at $x = 29.5a$ in a domain filled with epoxy and in two PCs of cadmium/epoxy and niobium/epoxy. All curves are normalized to the maximum value of displacement at the focal spot of the GRIN PC. The amplitude drop caused by solid cylinders in regular PCs and the GRIN PC is about 40% of the displacement amplitude in epoxy; this coincides with the reflection coefficient calculated from effective acoustic impedances. The focusing capability of the GRIN PC is evident from the magnification and distribution of the displacement field along the y axis [black line in Fig. 4.8(b)].

Fig. 4.9 shows the simulated displacement field of acoustic wave propagation in a

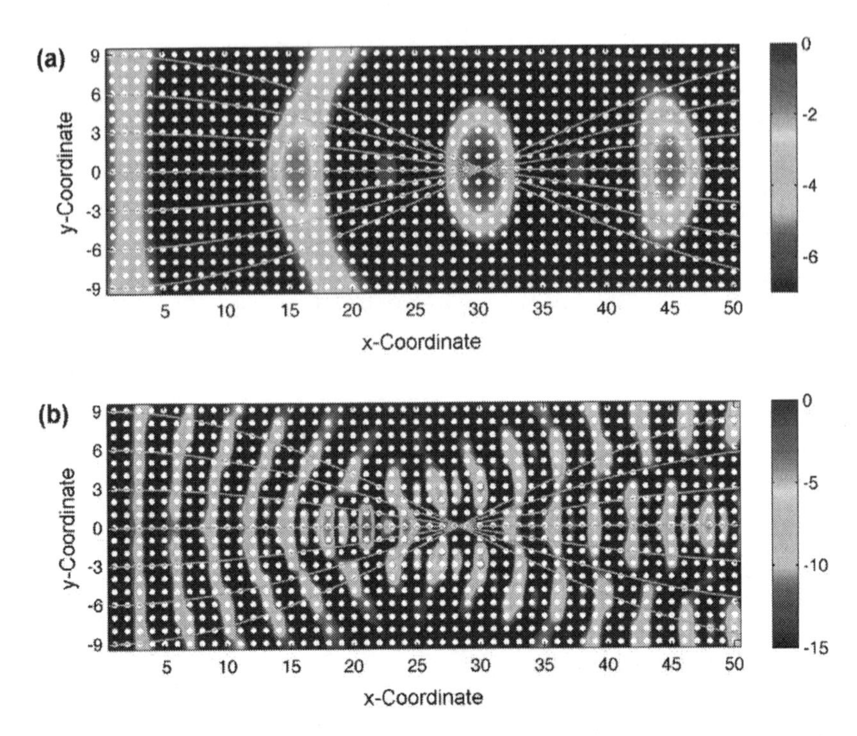

Figure 4.6. FDTD-simulated acoustic wave propagation in the GRIN PC at a reduced frequency of (a) $\Omega = 0.05$ and (b) $\Omega = 0.20$. The overlapped beam trajectories are calculated from Eq. 4.2. The amplitude of displacement fields are normalized and presented in decibel scale. The x- and y coordinates are in unit of a.

domain filled with epoxy and a 20-layer-thick GRIN PC at reduced frequency $\Omega = 0.10$. The planar acoustic waves are redirected within the GRIN PC and are focused outside the structure at $x = 29a$. These results indicate that GRIN PCs can serve as flat lenses that focus acoustic waves outside the structure. The focal spot is slightly longer and wider than that in Fig. 4.8(a) due to insufficient converging length. Nevertheless, we realized a clear shape of focal spot with -3 dB width of 60% of the wavelength.

The beam trajectories [Figs. 4.6(a), 4.6(b), 4.8(a) and 4.9] which overlapped on the simulated displacement field are calculated from the analytic expression in Sec. 4.2. For

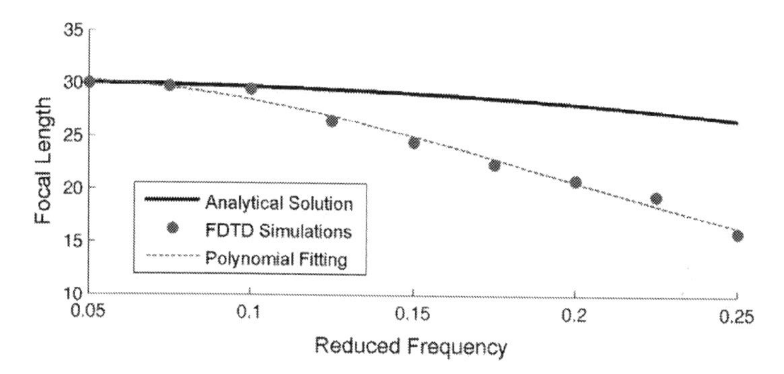

Figure 4.7. Calculated (solid line) and simulated (dotted line) focal lengths of the GRIN PC versus the operation frequency. The focal length is in unit of a.

trajectories outside the GRIN PC, the slopes of these straight rays are calculated by taking the derivative of Eq. 4.2 at the interface ($x = 20a$) and considering Snell's law of refraction. The focal spot predicted by the ray-tracing calculation matches well with the FDTD simulation, as indicated in Figs. 4.6(a), 4.8(a) and 4.9. We concluded that the analytic expression can serve as a powerful tool in designing flat GRIN PC lenses for acoustic focusing when the wavelength in the background material is much larger than the lattice spacing of the GRIN PC. Compared to existing negative-refraction-based PC lenses, our GRIN PC lens possesses several advantages. Firstly, the GRIN PC lens can operate over a wide frequency band below the first partial band gap; while negative-refraction-based PC lenses usually operate within a small range of partial band gaps. Secondly, a GRIN PC lens can be coupled with acoustic transducers and can effectively redirect paraxial incident acoustic waves to a small focal spot—the position of this spot is determined by the adjustable gradient coefficient. In contrast, with negative-refraction-based PC lenses one must focus select diverging waves to a long focal zone [18]. Finally, a GRIN PC lens can work at a reduced frequency range below $\Omega = 0.10$, so it can be made much smaller than a negative-refraction-based PC lens and can be seamlessly integrated with existing millimeter scale acoustic systems.

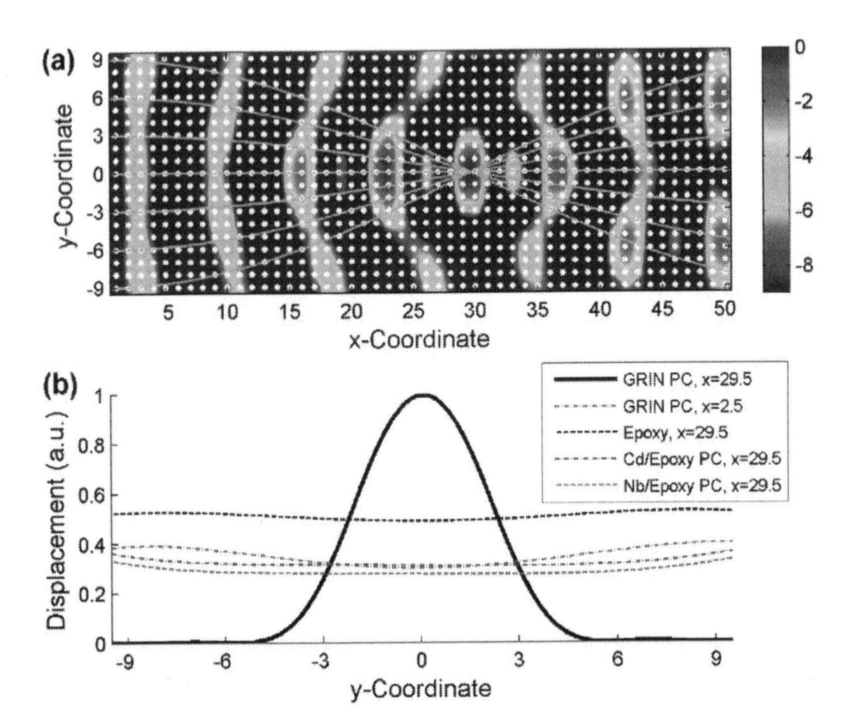

Figure 4.8. (a) FDTD-simulated acoustic wave propagation in the GRIN PC at a reduced frequency of $\Omega = 0.10$. (b) A comparison of displacement fields along y section of the GRIN PC, epoxy, and two regular PCs. All curves are normalized to the maximum displacement at the focal spot of the GRIN PC.

4.4 Summary

In this chapter, two-dimensional GRIN PCs with a hyperbolic secant gradient profile have been proposed to manipulate the propagation of acoustic waves. We numerically demonstrated that such a GRIN PC causes acoustic waves to converge over a wide range at low reduced frequencies (Ω =0.05–0.25), a range five times broader than those offered by existing negative-refraction-based PCs. The analytically calculated beam trajectories match well with FDTD-simulated results. Being a compact, flat acoustic lens with a

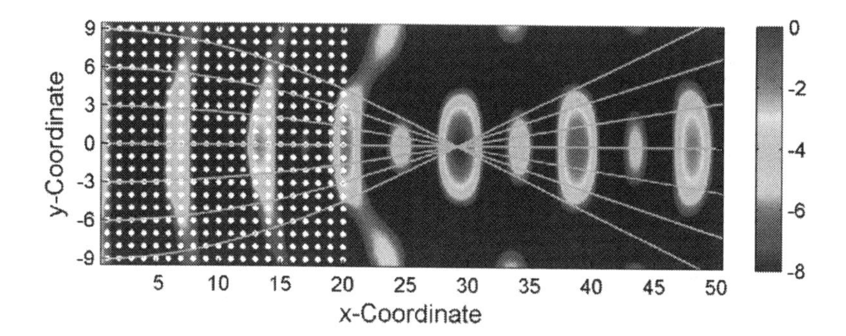

Figure 4.9. FDTD-simulated acoustic focusing in a domain filled with epoxy and a 20-layer-thick GRIN PC lens at a reduced frequency of $\Omega = 0.10$.

wide operational frequency range, the GRIN PC described in this chapter can be useful in applications such as acoustic imaging, drug delivery, high intensity focused ultrasound (HIFU), cell sonoporation, and non-destructive evaluation (NDE).

4.5 Further Reading - Experimental Verification

We are the first group introduced the GRIN PC concept to the phononic crystal community. Since then, several studies have been published by different groups to experimentally demonstrate the focusing ability of GRIN PCs. In 2010, the Sanchez-Dehesa group in Spain designed a two-dimensional GRIN PC that works at the first dispersion band. The design concept is the same as the one introduced in this chapter. However, the hyperbolic secant profile of their GRIN PC was obtained by modifying the radius of metal rods [57]. As shown in Fig. 4.10(A), a nine-layer GRIN PC was made of aluminum cylinders distributed in a hexagonal lattice with a lattice constant of $a = 2$ cm. The largest and smallest radii of cylinders used in this GRIN PC are 18 mm and 8 mm, respectively. Fig. 4.10(B) shows the sound amplification map for the GRIN PC at a frequency of 4.5 kHz (corresponding wavelength $\lambda \approx 3.8a = 7.6cm$). The focal spot has an amplification of 2 and a full width at half maximum (in the y direction) of 6 cm, equivalent to 80% of the wavelength. Same year, in collaboration with Dr. Sanchez-Dehesa, the Acoustics Division in United State Naval Research Lab fabricated a two-dimensional GRIN PC for aqueous

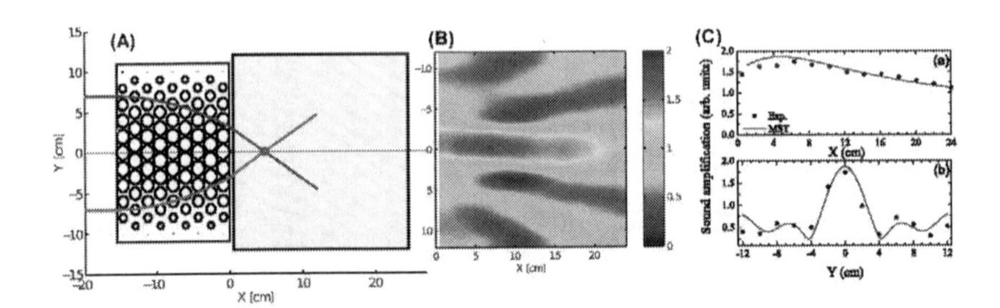

Figure 4.10. (A) Schematic of the GRIN PC in Sanchez-Dehesa's paper [57]. (B) Sound amplification map measured in the experiment at 4.5 kHz. (C) Longitudinal and transverse cuts extracted from (B). (Image source: doi: 10.1063/1.3488349)

applications [22]. The GRIN PC was designed in a similar fashion as in Fig. 4.10(A), and the amplification factor at the focal spot is 3 when operated at 20 kHz underwater.

In 2010, the Liu group in China experimentally observed the focusing effect for acoustic plane wave normally incident onto a two-dimensional GRIN PCs with negative gradient refractive index. As shown in Fig. 4.11(a), their GRIN PC is composed of steel rods (with a radius of 0.5 mm) arranged in a square-like array (with a lattice constant of 1.5 mm) and immersed in water. The gradient profile was achieved by gradual modification of the lattice constant along both longitudinal and transverse directions. It was designed to work at the second dispersion band in order to achieve negative refractive index for greater control on acoustic waves [58], similar to the method introduced in Chapter 2 to obtain large acoustic beam bending for acoustic mirage. The experimental results show the amplification factor of their GRIN PC is about 1.4 times to the input acoustic wave at the focal spot, and the full width at half maximum of the focus is about one wavelength [Fig. 4.11(b)].

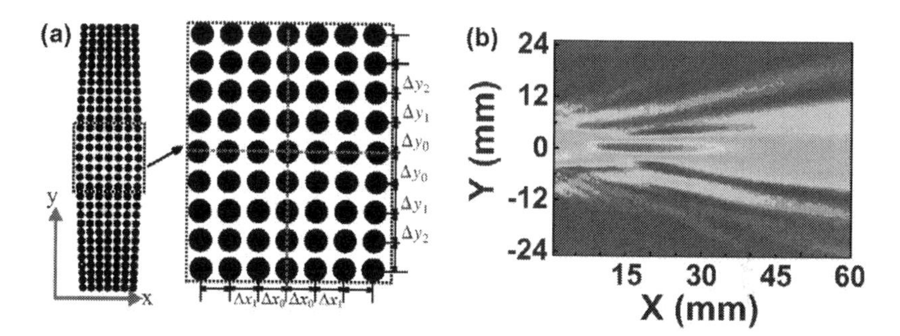

Figure 4.11. (a) Schematic of the GRIN PC in Liu's paper [58]. (b) Measured pressure field distribution of output wave at 0.75 MHz. (Image source: doi: 10.1063/1.3457447)

Chapter 5

Design of Acoustic Beamwidth Compressor Using Gradient-Index Phononic Crystals

In this chapter, we report a novel approach to effectively couple acoustic energy into a two-dimensional PC waveguide by an acoustic beamwidth compressor using the concept of hyperbolic-secant GRIN PCs. This chapter is organized as following. Section 5.1 describes the need for an acoustic beamwidth compressor and how a previously introduced hyperbolic-secant GRIN PC can fulfill this goal. Section 5.2 details the design of an acoustic beamwidth compressor for SV-mode BAW using the concept of a hyperbolic-secant GRIN PC. The hyperbolic-secant gradient profile is obtained by carefully choosing proper materials along the direction transverse to the acoustic wave propagation. Numerical investigation shows a beam-size conversion ratio of 6.5:1 and a transmission efficiency of up to 90% is obtained over the working frequency range of the PC waveguide. In Section 5.3, the concept is extended to PC plates to form an acoustic beamwidth compressor for the lowest antisymmetric Lamb wave. The hyperbolic-secant gradient profile is obtained by modulating the cylinder radius. Section 5.4 gives a summary of the chapter. Potential applications for this type of acoustic device include acoustic biosensors and signal processors. The work presented in this chapter has been featured as the front cover image in *Journal of Physics D: Applied Physics* [59] and reported in *Applied Physics Letters* [60].

5.1 Motivation

Recall that within phononic band gaps, any defect inside a PC will limit acoustic waves to propagate only within the defect. In the past years, researchers have developed PC-based waveguides that can efficiently guide acoustic waves along predefined pathways and sharp corners in a compact space on the order of several wavelengths down to nanometer scale [28, 30, 31, 25, 32, 33, 35]. The width of a PC waveguide is normally as narrow as one period of the crystal, allowing for wave propagation up to GHz range, which can prove useful in micro/nanoscale acoustic circuits. However, the dimension mismatch between the input acoustic beam and the PC waveguide could cause a great deal of energy loss, which is undesirable for practical applications. To fully realize the functionalities of a PC waveguide, it is necessary to effectively couple acoustic energy into the waveguide over its working bandwidth. Traditional acoustic lenses that focus acoustic waves through curved surfaces are prone to misalignment and difficult fabrication. While PC-based flat lenses using negative refraction could be conveniently integrated with PC waveguides, they are limited by their narrow working band and low gain.

To achieve efficient acoustic beamwidth compression, we propose a novel approach to concentrate acoustic waves by employing GRIN PCs with a hyperbolic secant profile. Using this approach, a wide acoustic beam can be effectively compressed and efficiently coupled into a narrow PC waveguide over a working bandwidth. This technique should be welcomed in applications where acoustic beamwidth compressors are needed to focus input and reference waves into a waveguide for nonlinear convolution processes [61].

5.2 Beamwidth Compressor for SV-BAW

In this section, we present the construction and characterization of a GRIN PC-based beamwidth compressor for SV-mode BAW.

5.2.1 Design of the Acoustic Beamwidth Compressor

The schematic illustration of the PC-based acoustic coupler is presented in Fig. 5.1. Although a PC supports propagation of multiple acoustic modes, in this chapter we focus our investigation on the SV mode BAW. As shown in Fig. 5.1, a 336 mm wide line source is placed at $x = 0$ and polarized along the z coordinate to generate a SV-mode BAW with

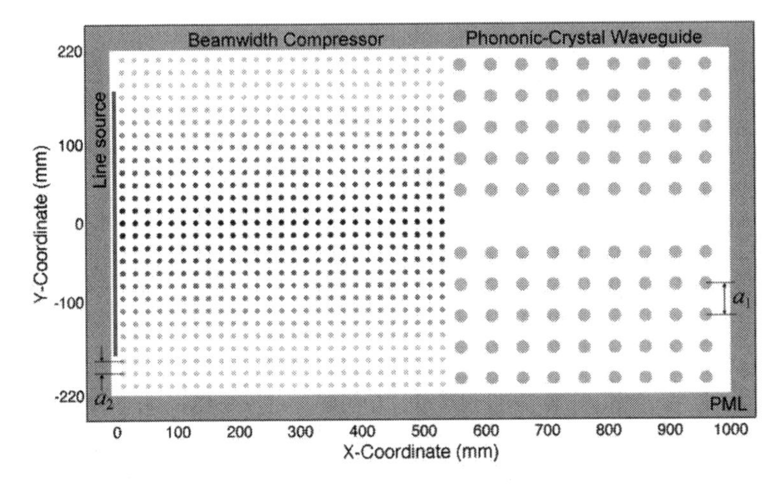

Figure 5.1. Schematic representation of the acoustic coupler design composed of a GRIN PC-based beamwidth compressor and a line-defect PC waveguide. Different colors represent different cylinder materials that are listed in Table 5.1. A line source is placed in front of the beamwidth compressor to excite SV-mode BAW.

a uniform-amplitude distribution across the y coordinate. A GRIN PC-based beamwidth compressor is introduced next to the line source to compress wide-beam plane waves to a focal spot at the inlet of a straight PC waveguide, guiding the coupled acoustic energy toward the outlet along the predefined path.

5.2.1.1 PC waveguide

The PC waveguide is obtained by simply removing the center row of cylinders along the wave propagation direction (ΓX orientation in the wave-vector space) from a square-lattice PC composed of gold cylinders embedded in epoxy, as shown in Fig. 5.1. The lattice period (a_1) and the gold cylinder radius (r_1) are 40 mm and 4 mm, respectively. The width of the PC waveguide, defined as the distance between adjacent cylinders on both sides of the waveguide, is 72 mm and its length is 9 periods or 360 mm. For this arrangement, the gold/epoxy PC exhibits a band gap for the SV-mode BAW from 10.88 (M_1) to 15.12 (X_2) kHz. Figure 5.2 shows the calculated band structures via a FDTD

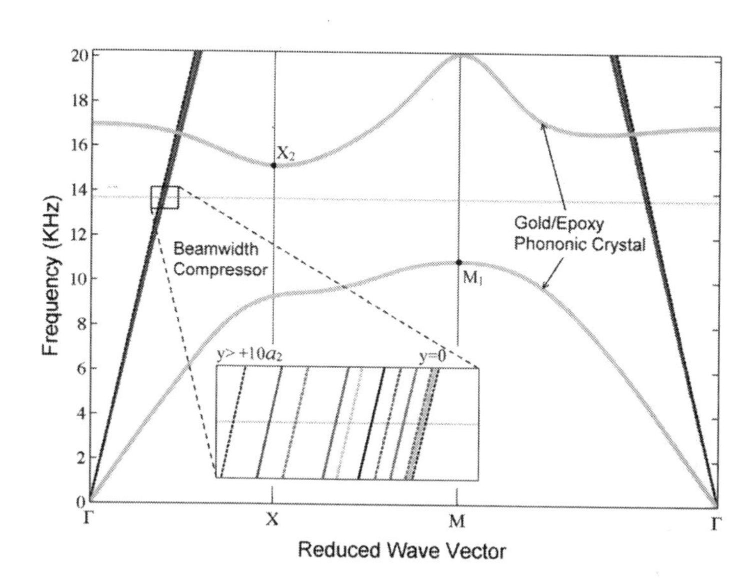

Figure 5.2. Band structure of the perfect gold/epoxy PC and each row of the GRIN PC-based beamwidth compressor for the SV-mode BAW. A hyperbolic secant gradient of acoustic velocity along the y coordinate in the beamwidth compressor is optimized at 13.66 KHz (doted horizontal line) to compress incident waves to the inlet of the PC waveguide.

method without considering the thermal effects on material properties. The material properties of gold and epoxy are listed in Table 5.1. Information about how to use the FDTD method to calculate the band structure of a PC or a PC waveguide is provided in Appendix B. In order to understand the guided modes associated with the straight PC waveguide, we calculate the dispersion curves of the waveguide by defining a supercell of 11 periods in the y coordinate in the FDTD calculation. As shown in Fig. 5.3, the band gap for the perfect gold/epoxy PC is delineated by two shaded regions in the ΓX orientation. Only one guided mode (or called defect mode) for the SV-mode BAW exists, extending from 10.88 to 14.88 kHz. A narrow guided-mode band gap is observed within the original phononic band gap of the perfect gold/epoxy PC structure. Hence, the PC waveguide can operate continuously over the frequency from 10.88 to 14.88 kHz.

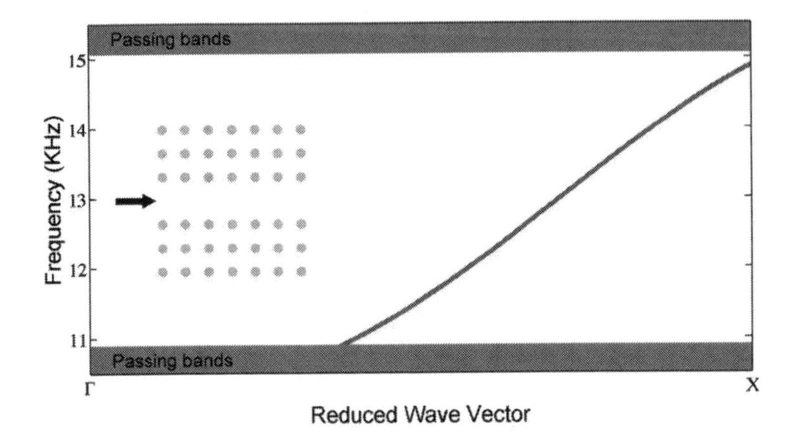

Figure 5.3. Dispersion curves of the guided modes of the PC waveguide.

5.2.1.2 Beamwidth compressor

In Chapter 4 we have introduced two-dimensional GRIN PC with a hyperbolic secant gradient profile. Through gradually modulating the material properties or filling fraction of a GRIN PC, one can induce an acoustic velocity variation along the modulating direction. Subwavelength focusing of planar acoustic waves by hyperbolic-secant GRIN PCs has been theoretically and experimentally demonstrated. Since the acoustic velocity in each row may vary for different propagating modes, GRIN PCs will have optimized performance for the designed acoustic mode. Based on this concept, the beamwidth compressor used in this chapter contains a 27×27 square array of solid cylinders embedded in epoxy. The lattice period (a_2) and the cylinder radius (r_2) are 16 mm and 1.6 mm, respectively. As shown in the schematic drawing in Fig. 5.1, 11 different solids are chosen to construct a hyperbolic secant acoustic velocity gradient along the y coordinate in the beamwidth compressor for the SV-mode BAW. The elastic properties of the chosen solids are listed in Table 5.1. The band structures of each row of the beamwidth compressor are calculated by the PWE method and plotted in Fig. 5.2. An enlargement at the design frequency of 13.66 kHz is shown in the insert of Fig. 5.2, from where the acoustic velocity in each row is determined from the relation: $v_g = \partial\omega/\partial k$, when ω represents angular velocity and k

Material	y coordinate in beamwidth compressor	Density (g/cm^3)	V_l (km/s)	V_s (km/s)
Epoxy		1.14	2.55	1.14
Gold		19.3	3.20	1.20
Copper	0	8.96	4.66	2.24
Bronze	$\pm a_2$	8.86	3.53	2.23
Niobium	$\pm 2a_2$	8.57	4.92	2.10
Titanium carbide	$\pm 3a_2$	10.15	6.70	2.96
Nickel	$\pm 4a_2$	8.84	5.70	2.24
Cast iron	$\pm 5a_2$	7.70	5.90	2.40
Zinc	$\pm 6a_2$	7.10	4.20	2.41
Zirconium	$\pm 7a_2$	6.48	4.65	2.25
Vanadium	$\pm 8a_2$	6.03	6.00	2.78
Nylon	$\pm 9a_2$	1.12	2.60	1.10
Titanium	$\pm 10a_2$	4.54	6.10	3.12

Table 5.1. Elastic properties of the materials used in the beamwidth compressor and PC waveguide.

represents the wave vector. Note that within the working frequency of the PC waveguide (10.88–14.88 kHz), the dispersion curves of all rows in both ΓX and ΓM orientations are nearly straight lines and thus the frequency-sensitivity of acoustic velocities is negligible. This fact suggests that though optimized at 13.66 kHz, the beamwidth compressor is very less frequency-dispersive and can focus acoustic waves over the frequency range in a stable manner.

The length of the beamwidth compressor used in the acoustic coupler design is determined from studying the propagation of acoustic waves in an independent GRIN PC. A GRIN PC with a length of 960 mm (60 periods) surrounded by 40-grid-thick perfectly matched layers at the boundaries is studied using the FDTD method. Figure 5.4 displays the simulated propagation of SV-BAWs excited at 13.66 kHz by a line source as described in Fig. 5.1. The amplitudes of displacement fields are shown in decibels (dB) and the color distribution is normalized to the maximum value of the displacement. From Fig. 5.4, wide-beam plane waves bend gradually toward $y = 0$ where acoustic velocity is lowest per Snell's law and converged at a focal zone centered at $x = 480$ mm (30 periods) due to the gradient. Beyond that zone, the focused beams are redirected parallel to the direction of propagation and reform planar wavefronts at $x = 960$ mm. It can be seen that peak acoustic intensity at the focal zone is 8 dB stronger than the input acoustic beam.

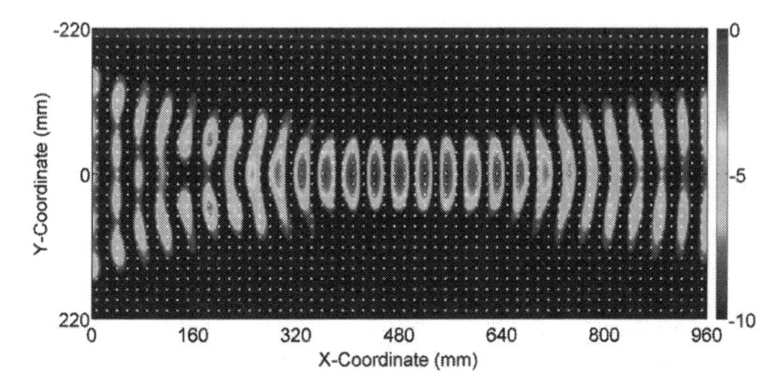

Figure 5.4. FDTD simulated acoustic wave propagation in a GRIN PC of length 960 mm (60 periods) at the design frequency of 13.66 KHz. The amplitudes of displacement fields are shown in decibels (dB).

The full-width at half maximum (FWHM) of the focused beam is 81 mm, equivalent to one wavelength of SV wave in epoxy (λ = 83.3 mm at 13.66 kHz). Since a beamwidth compressor of 27 periods (432 mm) is long enough for effective focusing, this length (432 mm) is used in the design of the acoustic coupler.

5.2.2 Results and Discussion

The FDTD-simulated SV-BAW propagation in the proposed PC-based acoustic coupler at 13.66 kHz is shown in Fig. 5.5. It can be seen that wide-beam acoustic waves are well compressed by the beamwidth compressor and coupled into the PC waveguide. To quantitatively analyze the effectiveness of the proposed beamwidth compressor, the steady-state transmitted acoustic energy is calculated at the outlet of the PC waveguide. Since half of acoustic energy generated by the line source propagates in the negative x direction, an intrinsic loss of -3 dB is included in the calculation of the transmission. The transmission spectra shown in Fig. 5.6(a) displays the transmissions of the acoustic coupler and a stand-alone PC waveguide where the same acoustic waves are directly fed into the structure. The transmission in both cases drops noticeably as frequency approaches the guided-mode band gap of the PC waveguide. The stand-alone PC waveguide exhibits a maximum transmission of -12.05 dB at 13.45 kHz, corresponding to a transmission rate

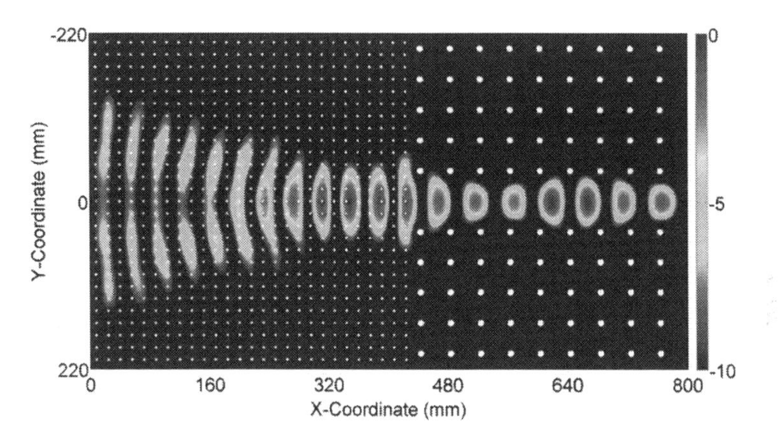

Figure 5.5. FDTD simulated acoustic wave propagation in the acoustic coupler at 13.66 KHz. Displacement fields are normalized to the maximum displacement of the guided beam.

of 12.5% (only considering acoustic energy propagating toward positive x direction). By comparison, the proposed acoustic coupler exhibits an average transmission of -3.85 dB over a broad range of frequencies from 10.88 to 14.50 kHz corresponding to a transmission rate of 82.5%. A transmission rate up to 90% is obtained at frequencies close to the design frequency of the beamwidth compressor. Thus, the beamwidth compressor contributes to a transmission enhancement of 7.2 times.

Figure 5.6(b) plots the transverse displacement profiles of the input and guided beams at the entrance of the beamwidth compressor and the outlet of the PC waveguide at 13.66 kHz. The profiles are normalized to the maximum amplitude of the guided beam. The input and guided beam has a FWHM of 336 mm and 51.5 mm, respectively. A beam-size conversion ratio of 6.5:1 is obtained in this work. While a reduction of the width of the PC waveguide can increase the conversion ratio, the inlet of the PC waveguide will need to be optimized to maintain good transmission. By increasing the number of rows in the beamwidth compressor and designing a smoother acoustic velocity gradient, wider acoustic beams can be compressed into the focal zone, increasing the conversion ratio without losing transmission efficiency.

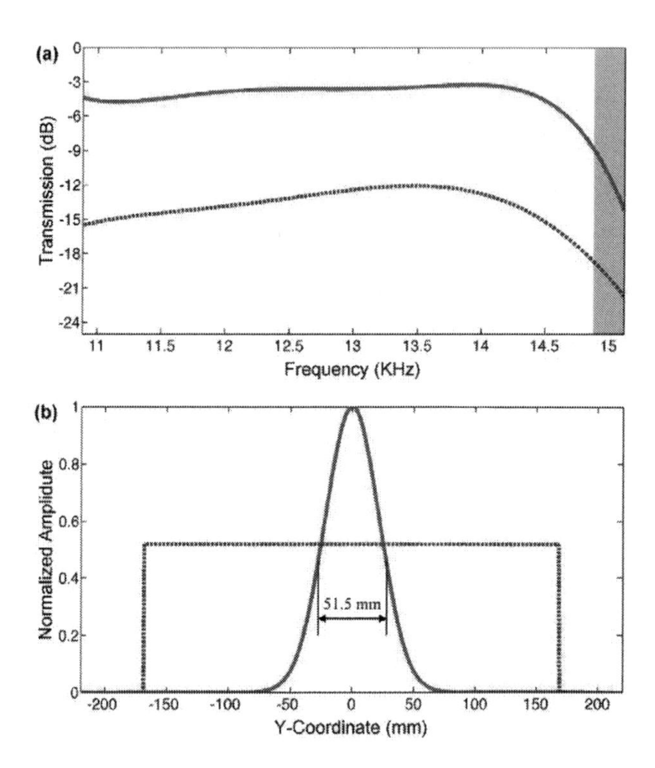

Figure 5.6. (a) Transmission for the acoustic coupler (red solid line) and a stand-alone PC waveguide (blue dashed line). The shaded region denotes the guided-mode bandgap of the PC waveguide. (b) The transverse displacement profiles of acoustic waves at the entrance of the beamwidth compressor (blue dashed line) and the outlet of the PC waveguide (red solid line) at 13.66 KHz.

5.3 Beamwidth Compressor for Lamb Waves

In addition to bulk PC waveguides, PC plate waveguides with a finite thickness have been shown to support Lamb wave propagation within the extended defects [11, 62, 16, 63]. The existence of phononic band gaps for Lamb waves depends strongly on not only the material properties and filling fractions but also the ratio of the plate thickness to the crystal periodicity. In fact, to construct PC plate waveguides is more practical since unbounded bulk PC waveguides are difficult to realize and unable to confine energy in the out-of-plane direction that leads to energy loss. Though promising, same dimension mismatch problem between the input acoustic beam and the PC plate waveguide as mentioned before could cause a great deal of energy loss, which is undesirable for practical applications. In this section, we demonstrate focusing of the lowest antisymmetric Lamb wave in a perforated GRIN PC silicon plate and its application as a beamwidth compressor.

5.3.1 Design of the Acoustic Beamwidth Compressor

Consider a perforated GRIN PC plate containing 15 rows of air holes arranged in a square lattice with graded filling fractions, as shown in Fig. 5.7(a). The thickness of the PC plate is 50 μm while the lattice constant is $a = 100$ μm. Fig. 5.7(b) shows band structure for the lowest antisymmetric Lamb wave (A_0 mode) of a perfect-periodic PC plate with different filling fractions. We observe that the first dispersion band drops and the group velocity ($v_g = \partial\omega/\partial k$) decreases with the increase of the filling fraction. Fig. 5.7(c) shows the equal-frequency contours of the A_0 mode Lamb wave calculated at the design frequency of 3 MHz with various filling fractions. For filling fractions smaller than 0.503, the equal-frequency contours are close to circles, implying that anisotropy of either the phase and group velocities are small enough to be treated as weakly anisotropic. Therefore, we could obtain a hyperbolic secant gradient profile in the GRIN PC plate by gradually modulating the radius of the air holes along the transverse direction (y axis), as shown in Fig. 5.7(a).

The radii of the air holes at the center row ($y = 0$) and the boundary rows ($y = \pm 7a$) of the GRIN PC plate are assumed to be $r_0 = 40$ μm and $r_7 = 20$ μm, respectively. Using Eq. (4.6), we calculate the corresponding effective refractive indices as 1.159 and 1.044, respectively. By fitting a hyperbolic secant profile, the gradient coefficient and radius of air holes at each row of the GRIN PC plate can be determined. The focusing of the A_0 mode Lamb wave in the designed GRIN PC plate is shown in Fig. 5.7(a). Planar Lamb waves excited at $x = -2a$ from a line source are focused at a long focal zone extending from $x = 20a$ to $x = 32a$ with a maximum amplification factor of 2.13. To test the feasibility of compressing the wave beam of a plate wave into a PC plate waveguide with small aperture, the GRIN PC plate with a thickness of 50 μm is trimmed at $x = 22a$ and utilized as a beamwidth compressor.

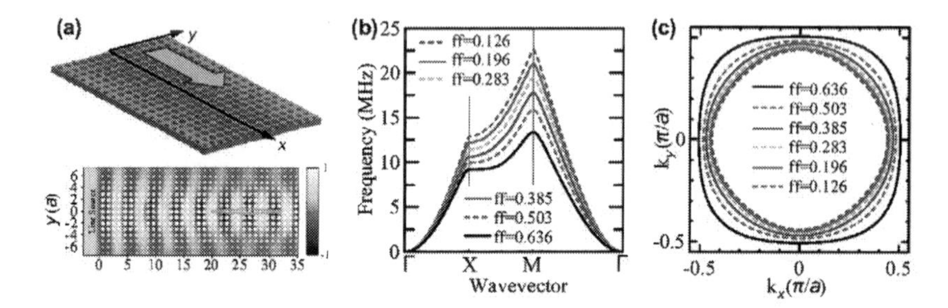

Figure 5.7. (a) Schematic of the GRIN PC plate used to couple Lamb wave to a PC plate waveguide. Band structure of the A_0 mode of the air/silicon PC plate with different filling fractions. (b) The equal-frequency contours of the A_0 mode at 3 MHz.

5.3.2 Coupling efficiency to a PC plate waveguide

A PC plate waveguide is constructed by removing two rows of cylinders in the x direction from a PC plate composed of periodic stubbed tungsten cylinders on a silicon plate. The thickness of the silicon plate is the same as that of the GRIN PC plate (50 μm). By choosing the lattice constant of the PC plate waveguide as $a = 150$ μm and the radius and height of the tungsten cylinders as 36 μm and 273 μm, respectively, a complete phononic band gap for the A_0-mode Lamb wave can be found in the range of 2.6–3.4 MHz. The PC plate waveguide is placed as an adjunct to the GRIN PC plate, as shown in Fig. 5.8.

Fig. 5.8(a) shows the simulated Lamb wave propagation for A_0 mode at the design frequency 3 MHz. It can be clearly seen that planar waves with a width of $14a$ are compressed by the GRIN PC plate and guided into the PC plate waveguide. The results show that the amplitude of the displacement field at the entrance of the waveguide ($x = 22a$) is amplified by a factor of 3.2 times to that of the input waves. The full-width at half maximum of the compression rate of about 5.8:1. To demonstrate the proposed GRIN PC plate can efficiently couple acoustic waves to the PC plate waveguide over a wide frequency band, numerical simulations with operating frequency at 2.7 MHz and 3.4 MHz are also conducted. Fig. 5.8(b) shows the simulated Lamb wave propagation for A_0 mode at 2.7 MHz. The simulation results demonstrate that the amplification factor at the entrance of the PC plate waveguide is 2.5 and 2.6 for 2.7 MHz and 3.4 MHz, respectively, indicating good coupling efficiency. Hence we prove that the proposed GRIN PC plate

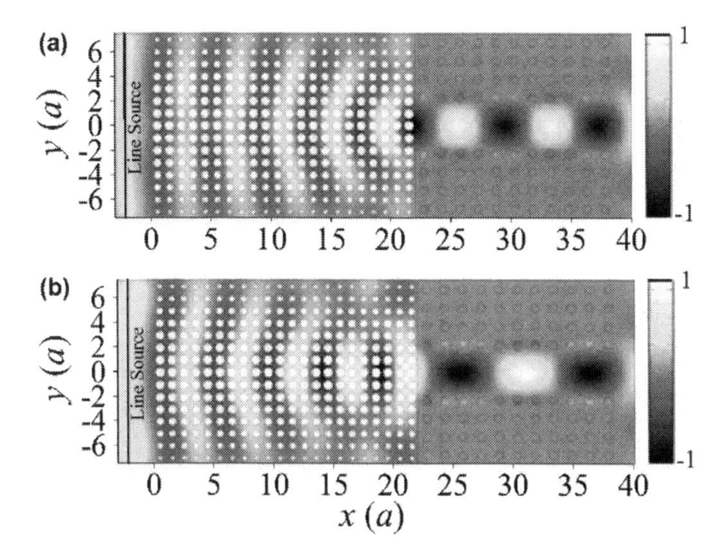

Figure 5.8. Simulated wave propagation along the x direction in a GRIN PC plate adjoined with a PC plate waveguide at (a) 3 MHz and (b) 2.7 Hz.

based acoustic beamwidth compressor can work efficiently over a wide band width of 23% of the design frequency.

5.4 Summary

In this chapter, we demonstrate two beamwidth compressors using the concept of GRIN PC to couple acoustic waves into PC waveguides. The GRIN PC-based beamwidth compressor for SV-BAW exhibits a beam-size conversion ratio of 6.5:1 and a transmission efficiency up to 90% over the working frequency range of a straight PC waveguide. A transmission enhancement of 7.2 times is obtained compared to a stand-alone PC waveguide. The GRIN PC plate-based beamwidth compressor for lowest antisymmetric Lamb wave exhibits a beam-size conversion ratio of 5.8:1 and a displacement amplification of 3.2. The working frequency range is as wide as 23% of the design frequency. This type of acoustic beamwidth compressor is compact and flat, and it operates stably over a wide frequency range. Due to these advantages, the methodology described in this work can be useful in many applications relating to signal processors, acoustic sensors and cell patterning.

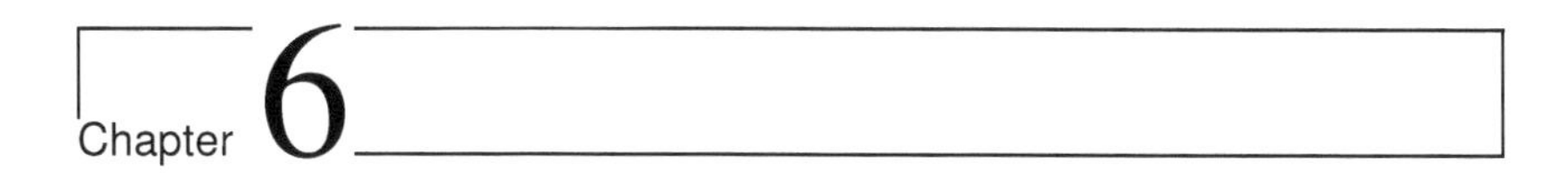

Design of Acoustic Beam Aperture Modifier Using Gradient-Index Phononic Crystals

In previous chapters, we have shown the versatility of devices that control acoustic wave propagation in a PC structure is enhanced by introducing the concept of gradient index. Focusing of planar acoustic waves has been successfully achieved by two-dimensional GRIN PCs whose hyperbolic-secant gradient profiles are obtained either by changing the material or by modulating the filling fraction row by row. To further explore the functionalities of GRIN PCs, we design a GRIN PC-based acoustic beam aperture modifier to efficiently vary an acoustic beam aperture, i.e. increase or decrease the beam aperture with minimum energy loss and wave form distortion. The design of a novel acoustic beam aperture modifier using hyperbolic-secant GRIN PCs consisting of same-size steel cylinders embedded in a homogeneous epoxy background is reported in Section 6.1. The hyperbolic secant gradient profiles of the GRIN PCs are obtained by modulating the lattice constant along the transverse direction to the phononic propagation. Section 6.2 presents simulated wave propagation in the proposed acoustic beam aperture modifiers. The results show that beam aperture can be modified while acoustic energy and waveform are well conserved by the proposed devices.

The concept presented here has been reported in *Journal of Applied Physics* [64] and won the Student Paper Competition Award in 2010 IEEE International Ultrasonic Symposium.

6.1 Design of Acoustic Beam Aperture Modifier

The proposed acoustic beam aperture modifier is composed of two butt-jointed hyperbolic-secant GRIN PCs, as portrayed in Fig. 6.1. Each GRIN PC is a two-dimensional periodic structure composed of infinite-length rigid steel cylinders arranged in a square lattice and embedded in epoxy. Two GRIN PCs are designed in a similar fashion, but with different hyperbolic-secant gradient profiles, $n_1(y)$ and $n_2(y)$. When planar acoustic waves are propagating in the positive x direction, the first GRIN PC (GRIN PC 1) serves as a converging lens to compress planar waves to a focal spot located at the interface of two GRIN PCs. The second GRIN PC then serve as a collimating lens to guide the focused acoustic waves back to planar and parallel to the x axis. The output acoustic beam has a different aperture than that of the input acoustic beam since the second GRIN PC (GRIN PC 2) has a sharper hyperbolic-secant refractive index profile. Likewise, the aperture of an acoustic beam propagating in the negative x direction will be expanded by the butt-jointed GRIN PCs. Therefore, convenient beam aperture modification can be obtained by the proposed device.

In Fig. 6.1, two GRIN PCs are filled with steel cylinders of equal radius and have same lattice constant a_x in the wave propagation direction (along x axis). The hyperbolic-secant refractive index profile in each GRIN PC is obtained by adjusting the lattice constant a_y in the transverse direction (along y axis). To explore the dependency of effective refractive index of a PC on its lateral lattice constant a_y, we calculate the band structures for the SV-mode BAW of square-lattice steel/epoxy PCs with different aspect ratios ($AR = a_y/a_x = h_i/w$) using the FDTD method (see Appendix B.2). The radius of steel cylinders is assumed to be $r = 0.2a$. The material properties of steel and epoxy are the same as used in Chapter 3. Fig. 6.2(a) shows the first two bands of the dispersion curves for SV-mode BAW propagating in six different PCs. Since the lattice constant a_x in the x direction is kept constant and only the lateral lattice constant a_y is adjusted, the change in dispersion curves with varying aspect ratios is different in ΓX and ΓM orientations. With an increase in aspect ratio, the first dispersion band of the PC drops and the group velocity decreases steadily along ΓX orientation, while it remains almost unchanged along ΓM orientation. The calculated refractive index along ΓX and ΓM orientations at a reduced frequency of 0.25 are plotted in Fig. 6.2(b). The dependence of the refractive index along ΓM orientation ($n_{\Gamma M}$) on aspect ratio is very linear, decreasing

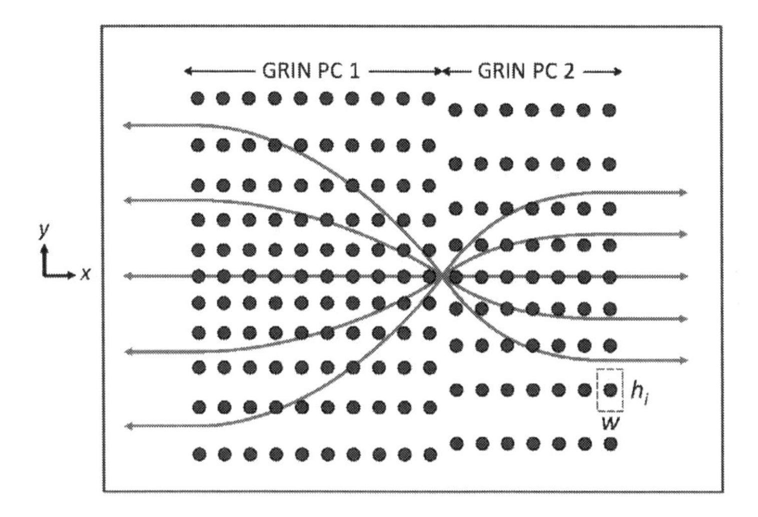

Figure 6.1. Schematic of the acoustic beam aperture modifier design composed of two butt-jointed two-dimensional GRIN PCs with different gradient profiles. The black dots represent the locations of steel cylinders, and the background material is epoxy. The aperture of a planar SV-mode BAW can be modified from wide to narrow or narrow to wide.

continuously from 1.25 to 1.13. On the contrary, the dependence of the refractive index along ΓX orientation ($n_{\Gamma X}$) on aspect ratio is nearly negligible. The effective refractive index n_{eff} can be calculated by taking the average of $n_{\Gamma X}$ and $n_{\Gamma M}$, as defined in Eq. (4.6).

The lateral lattice constant a_y at each row of the GRIN PC is carefully chosen to form a hyperbolic-secant refractive index profile across the structure. Fig. 6.3 shows the effective refractive index profiles of three different GRIN PCs, each can be fitted perfectly with a hyperbolic secant function. As a result, each GRIN PC can be used to focus planar acoustic waves or collimate acoustic waves emitted from a point source.

To verify the design, we simulate the acoustic wave propagation through GRIN PC 1 using the FDTD program at the design frequency 17.78 KHz (corresponding to the reduced frequency $\Omega = 0.25$). The simulation domain used for FDTD is a two-dimensional GRIN PC that comprises 25 rows in the y direction and 80 layers in the x direction. The lattice

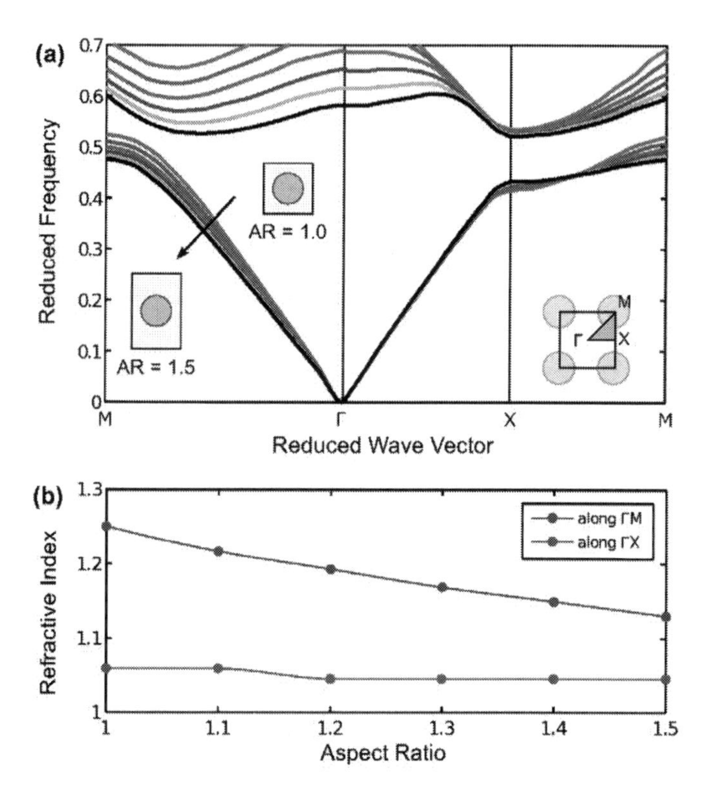

Figure 6.2. (a) Band structures of rectangular-latticed steel/epoxy PC with different aspect ratios for the SV-mode BAW. (b) Refractive index of the steel/epoxy PC along ΓX and ΓM orientations against aspect ratio at reduced frequency of 0.25.

spacing a_x in the x direction is set as 8 mm and the whole domain is discretized uniformly into a density of 20 spatial grids per 16 mm. A 400 mm long line source is placed at $x = 0$ to generate a planar SV-mode BAW. The entire domain is surrounded by 40-grid-thick perfectly matched layers at the boundaries to avoid numerical reflections in the simulation domain. As shown in Fig. 6.4, a full-width, normally incident acoustic beam converges to a focal spot at $y = 0$, $x = 38a = 608$ mm due to the gradient of GRIN PC 1. Beyond the focal spot, the focused beam is reconstructed to a planar wave. The overlapped beam

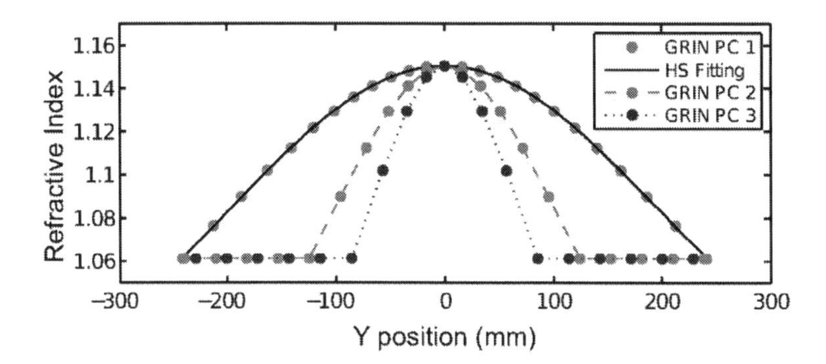

Figure 6.3. Effective refractive index profiles of three different GRIN PCs, each can be fitted perfectly with a hyperbolic secant function.

trajectories are plotted to help visualizing the wave formation.

Note that GRIN PCs can converge normal incident beam to a single spot, and vise versa. The sharper hyperbolic secant gradient is constructed in the GRIN PC, the shorter focal length is obtained. Hence, to design the acoustic beam aperture modifier we use GRIN PC 1 with a length of 608 mm ($38a$) to converge the full-width (400 mm wide) planar acoustic beam to the interface of two GRIN PCs, and then use GRIN PC 2 to recover the acoustic energy to planar waves with a narrower beam aperture, as shown in Fig. 6.1.

6.2 Results and Discussion

The acoustic beam aperture modification by the proposed device is numerically verified in this section. The FDTD-simulated wave propagation in the proposed GRIN PC-based acoustic beam aperture modifier at a normalized frequency of 0.25 is shown in Fig. 6.5. It can be seen from Fig. 6.5(a) that the full-width of the acoustic beam is well focused at the interface of GRIN PC 1 and GRIN PC 2 ($x = 608$ mm) and then transmitted into GRIN PC 2. No reflection is observed at the interface due to the first columns of GRIN PC 2 are positioned exactly at the periodic frame of GRIN PC 1. The acoustic intensity at the focused spot is around 7 dB stronger than that of the incident waves. The focal

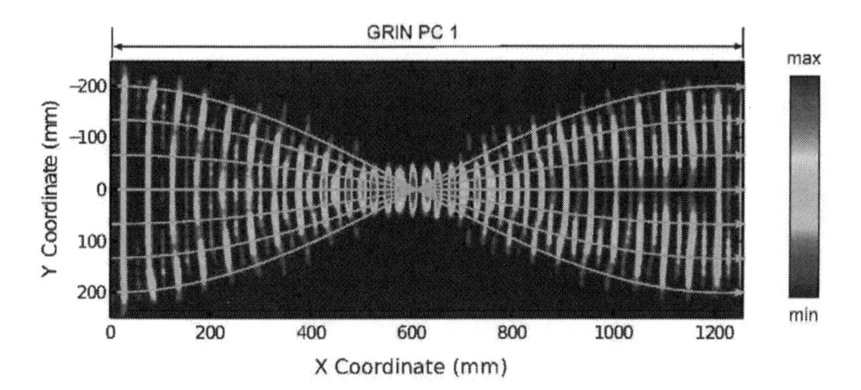

Figure 6.4. FDTD simulated acoustic wave propagation in GRIN PC 1 at the design frequency of 17.78 kHz ($\Omega = 0.25$). The line source width is 400 mm wide and the wave aperture at the output is maintained.

zone has a -3 dB width (in the y direction) of less than 54 mm ($3.4a$), which is 85% of the wavelength of SV wave in epoxy at this frequency , which is a little bigger than the wavelength of SV wave in epoxy at this frequency ($\lambda = a/\Omega = 4a$). After entering GRIN PC 2, the acoustic beam is gradually expanded and redirected to the direction parallel to x axis at $x = 912$ mm $= 57a$, which is exactly the sum of focal lengths of GRIN PC 1 ($38a$) and GRIN PC 2 ($19a$). After passing through GRIN PC 2, the transmitted acoustic beam is still collimated when propagating in epoxy. The output beam aperture at $x = 1060$ mm is 210 mm, which is close to half of the width of the line source (400 mm). Thus the proposed beam aperture modifier achieves a 52.5% beam aperture conversion with around 90% acoustic energy conserved.

A beam aperture modifier composed of GRIN PC 1 and GRIN PC 3 is shown in Fig. 6.5(b). The output beam aperture is measured as 140 mm, which is 35% wide of the input beam aperture. The output beam has conserved 83% of the acoustic energy and is well collimated, showing no sign of converging or diverging. This demonstrates the effectiveness of our proposed beam aperture modifier.

In this section, we only show the simulated wave propagation in the proposed device for the design frequency. However, due to the fact that the dispersion curves in Fig. 6.2 are nearly linear in the frequency range between 0.0 and 0.4, the proposed GRIN PC-

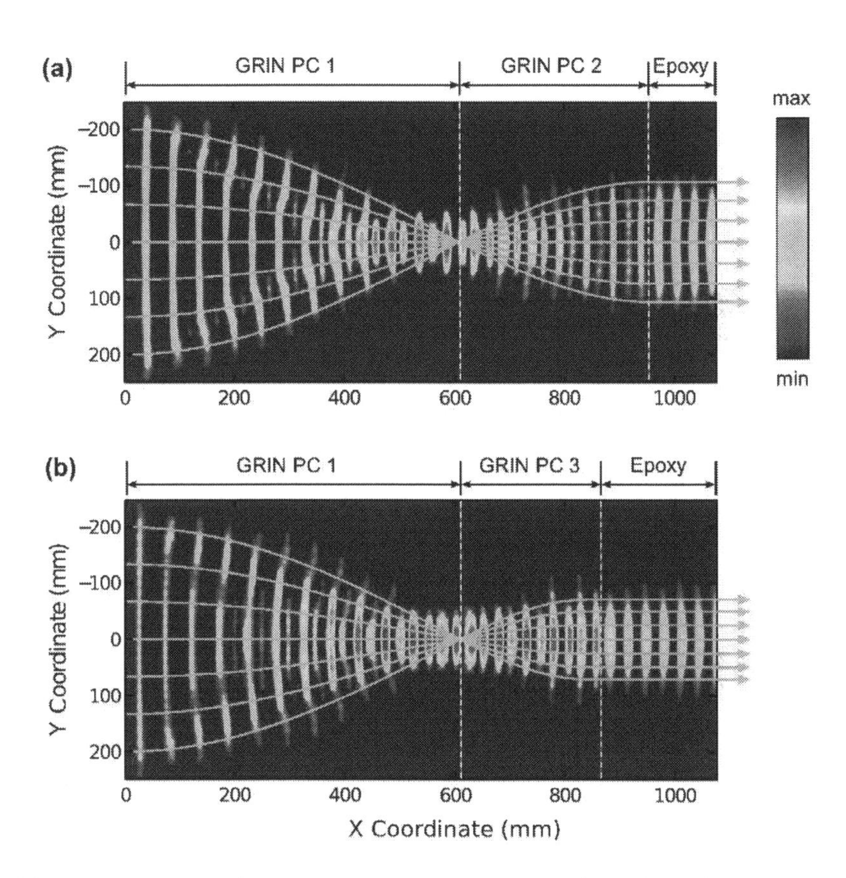

Figure 6.5. FDTD simulated acoustic wave propagation in (a) GRIN PC 1 + GRIN PC 2 and (b) GRIN PC 1 + GRIN PC 3 at the design frequency of 17.78 kHz. The line source width is 400 mm wide and the wave aperture at the output is (a) 210 mm and (b) 140 mm.

based aperture modifier is weakly frequency-dispersive and thus should operate over a wide frequency range without considerable degradation.

6.3 Summary

In this chapter, we computationally demonstrate a design of two-dimensional GRIN PC-based acoustic beam aperture modifier that has the ability to change the beam aperture while conserving acoustic energy. The beam aperture modification is obtained by adjusting the constitutive parameters of the GRIN PCs along the direction parallel to incident waves. The analytical solution and simulation are in good agreement. The device investigated in this work could be useful in applications relating to acoustic imaging and nondestructive evaluation.

Chapter 7

Tunable Phononic Crystals with Anisotropic Inclusions

In this chapter we present a theoretical study on the tunability of phononic band gaps in two-dimensional PCs consisting of various anisotropic cylinders in an isotropic host. A two-dimensional PWE method is used to analyze the band structure of the PCs; the anisotropic materials used in this work include cubic, hexagonal, trigonal, and tetragonal crystal systems. By reorienting the anisotropic cylinders, we show that phononic band gaps for bulk acoustic waves propagating in the PC can be opened, modulated, and closed. The methodology presented here enables enhanced control over acoustic metamaterials which have applications in ultrasonic imaging, acoustic therapy, and nondestructive evaluation. This chapter is organized as following. Section 7.1 describes the motivation for the study. In Section 7.2, we introduce the model and method used to calculate the band gaps of the PCs. In Section 7.3, we present and discuss the tunablility of two-dimensional PCs with cubic, hexagonal, trigonal, and tetragonal cylinders, respectively. Finally, a summary is given in Section 7.4. The work presented in this chapter has been reported in *Physical Review B* [65].

7.1 Motivation

Understanding the propagation behavior of acoustic waves in periodic composite structures is important for many acoustic-based applications. Much of the existing research on PCs focuses on understanding the dependence of the width and position of phononic band gaps upon the constitutive parameters (e.g. geometry, composition, and material properties) of the structures [44]. This focus is due to the fact that nearly all phenomena that a PC exhibits rely on the formation of a phononic band gap. Explicit control of a phononic band gaps yields desirable operation parameters and improves overall performance in PC-based applications. Recent studies have shown that phononic band gaps can be tuned by (i) physical rotation [43, 45, 66] or relocation [67] of inclusions to modify the band structure for acoustic waves, (ii) mechanical deformation of the structure by an external stress [47], (iii) actively changing the elastic properties of the constitutive materials through applications of strong external stimuli (e.g. electric and magnetic fields) [46, 48, 50], and (iv) changing the acoustic velocities in ferroelectric materials through a temperature variation induced phase transition [68, 69]. For cases (i) and (ii), isotropic materials are usually chosen to form the heterogeneous structures with a change in geometry accounting for the tunability of the phononic band gaps. On the contrary, a number of anisotropic materials have been utilized to build tunable PCs with approaches (iii) and (iv), as they are more sensitive to perturbations in the environment. Because of this fact, the rotation/relocation of anisotropic inclusions in a PC could potentially have greater flexibility and stronger effects in tuning phononic band gaps, however, no comprehensive research has been done yet.

In this chapter, we present a comprehensive study on the tunability of phononic band gaps in two-dimensional PCs consisting of anisotropic cylinders in an isotropic host, where control of the phononic band gaps results from the reorientation of the anisotropic cylinders. The anisotropic materials considered in this study include cubic, hexagonal, trigonal, and tetragonal materials. We use a two-dimensional PWE method to calculate the variation in phononic band gaps of bulk acoustic waves due to the reorientation of the anisotropic cylinders. Our results show that the band gaps of a PC can be tuned largely by the proposed method, allowing enhanced control over acoustic metamaterials.

7.2 Formulation

Figure 7.1 shows the schematic geometry of the periodic composite structures used in our work. The anisotropic cylinders with radius r are arranged in a two-dimensional square array with lattice spacing d and embedded in a homogeneous host material. A global Cartesian coordinate system xyz is set with the z axis parallel to the cylinder axis and

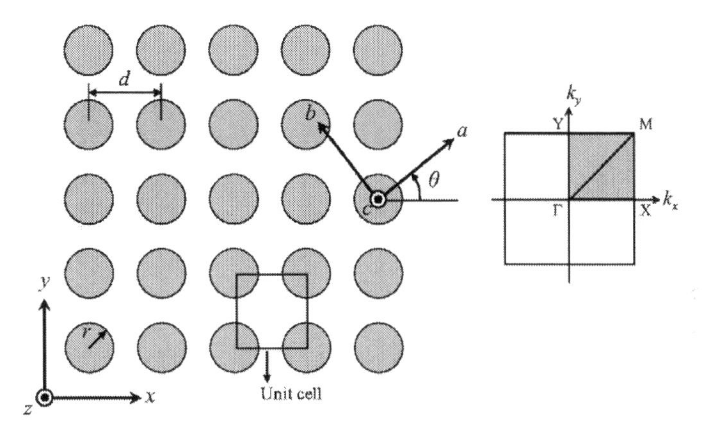

Figure 7.1. The left diagram shows the top view of an infinite two-dimensional square-lattice PC. The right diagram shows the corresponding first Brillouin zone of the unit cell.

x and y axes along the lattice axes of the square array. In the absence of a body force and temperature factor, the equation of motion for the displacement vector in a periodic composite structure can be written as

$$\rho(\mathbf{r})\ddot{u}_i(\mathbf{r},t) = \partial_j[c_{ijmn}(\mathbf{r})\partial_n u_m(\mathbf{r},t)], \tag{7.1}$$

where $\mathbf{r} = (\mathbf{x}, z) = (x, y, z)$ is the position vector with respect to the global coordinate system xyz, t is the time variable, $\rho(\mathbf{r})$ and $c_{ijmn}(\mathbf{r})$ are the position-dependent mass density and elastic stiffness tensor, respectively. In this section, the constitutive materials (cylinders and host) of the periodic composite structure are of the lowest symmetry, i.e., belonging to the triclinic symmetry.

The wave behavior in this two-dimensional PC is very complicated and highly sensitive to the frequency and direction of the propagating wave due to the periodicity of the structure and the anisotropy of the constitutive materials, . The wave equation, Eq. (7.1(, can be solved by a number of numerical methods, such as PWE methods, MST, and FDTD methods. In this chapter we use a two-dimensional PWE method to solve for the band structure of the acoustic waves propagating in the periodic composite structures because of its strength in analyzing structures containing of anisotropic materials. According to the PWE method, when only bulk acoustic modes are considered, the equation of motion for a two-dimensional periodic composite structure can be expanded into a Fourier series with respect to reciprocal lattice vectors $\mathbf{G}$ and $\mathbf{G}'$, and rewritten as an eigenvalue problem. The formation of the eigenvalue problem is detailed in Appendix A. Once the eigenfrequencies are obtained, the relative amplitude of the displacement for each eigenmode can be solved by substituting the eigenfrequencies into Eq. (A.2).

As the anisotropic cylinders are rotated through an angle θ about the cylinder axis (z axis), the crystalline axes of the cylinders are reoriented accordingly. The rotation of the cylinders changes the value of the Fourier coefficients of the elastic stiffness tensor $c^{ij}_{\mathbf{G}-\mathbf{G}'}$ in Eqs. (A.10)–(A.13) which influences the acoustic wave propagation behavior in the periodic composite structure. We define another set of Cartesian coordinate system abc with c axis parallel to the z axis of the global system, as shown in Fig. 7.1, to track the orientation of the anisotropic cylinders. The transformation of the coordinate position vector between the global and cylinder coordinate systems can be achieved by

$$\begin{bmatrix} a \\ b \\ c \end{bmatrix} = \begin{bmatrix} \cos\theta & \sin\theta & 0 \\ -\sin\theta & \cos\theta & 0 \\ 0 & 0 & 1 \end{bmatrix} \cdot \begin{bmatrix} x \\ y \\ z \end{bmatrix}. \tag{7.2}$$

Consequently, the elastic stiffness tensor of the rotated cylinders from the view of the global coordinate system xyz can be obtained by transforming the rotational coordinate [70]:

$$\mathbf{c}' = \mathbf{T}\mathbf{c}\mathbf{T}^t, \tag{7.3}$$

where

$$\mathbf{T} = \begin{bmatrix} \cos^2\theta & \sin^2\theta & 0 & 0 & 0 & \sin 2\theta \\ \sin^2\theta & \cos^2\theta & 0 & 0 & 0 & -\sin 2\theta \\ 0 & 0 & 1 & 0 & 0 & 0 \\ 0 & 0 & 0 & \cos\theta & -\sin\theta & 0 \\ 0 & 0 & 0 & \sin\theta & \cos\theta & 0 \\ -\sin\theta\cos\theta & \sin\theta\cos\theta & 0 & 0 & 0 & \cos 2\theta \end{bmatrix} \tag{7.4}$$

is the transformation matrix and $\mathbf{T}^t$ is the transpose of $\mathbf{T}$. Once the elastic stiffness tensor $c^{ij}_{\mathbf{G}-\mathbf{G}'}$ is obtained, the new band structure for the bulk acoustic modes can be calculated by substituting $\mathbf{c}'$ back into Eqs. (A.10)–(A.13) and solving the eigenvalue problem.

The elastic material properties used in our work are listed in Table 7.1. The elastic

Material	Symmetry	Density (kg/m^3)	C_{11}	C_{12}	C_{13}	C_{14}	C_{33}	C_{44}	C_{66}
			Elastic stiffness ($\times 10^{10}$ N/m^2)						
Epoxy	isotropic	1180	0.76					0.16	
GaAs	cubic	5307	11.9	5.38				5.94	
ZnO	hexagonal	5680	21.0	12.1	10.5		21.1	4.25	
Quartz	trigonal	2651	8.67	0.70	1.19	1.79	10.7	5.79	
TiO_2	tetragonal	4260	26.6	17.3	13.6		47.0	12.4	18.9

Table 7.1. Elastic properties of the materials used in this chapter.

stiffness components given in Table 7.1 refer to the global coordinate axes x, y, z that coincide with the crystalline axes X, Y, Z, i.e., acoustic waves propagate in the XY plane of the anisotropic materials. When acoustic waves do not propagate in the XY plane of the cylinders, i.e., the cylinder coordinate axes a, b, c do not coincide with the crystalline axes X, Y, Z, corresponding coordinate transformations must be applied to the elastic stiffness tensor of the cylinders before solving the eigenvalue problem. As acoustic waves propagate in the XZ plane of the cylinders, for example, the embedded crystal experiences a clockwise rotation through 90° about the crystalline X axis. Or when acoustic waves propagate in a meridian plane, the crystal rotates through 45° about the crystalline Z axis followed by 90° about the crystalline X axis. The coordinate transformation matrices for a rotation through an angle ξ about the crystalline X axis and an angle η about the crystalline Y axis are

$$\mathbf{T}_X = \begin{bmatrix} 1 & 0 & 0 & 0 & 0 & 0 \\ 0 & \cos^2\xi & \sin^2\xi & \sin 2\xi & 0 & 0 \\ 0 & \sin^2\xi & \cos^2\xi & -\sin 2\xi & 0 & 0 \\ 0 & -\sin\xi\cos\xi & \sin\xi\cos\xi & \cos 2\xi & 0 & 0 \\ 0 & 0 & 0 & 0 & \cos\xi & -\sin\xi \\ 0 & 0 & 0 & 0 & \sin\xi & \cos\xi \end{bmatrix} \tag{7.5}$$

and

$$\mathbf{T}_Y = \begin{bmatrix} \cos^2\eta & 0 & \sin^2\eta & 0 & -\sin 2\eta & 0 \\ 0 & 1 & 0 & 0 & 0 & 0 \\ \sin^2\eta & 0 & \cos^2\eta & 0 & \sin 2\eta & 0 \\ 0 & 0 & 0 & \cos\eta & 0 & \sin\eta \\ \sin\eta\cos\eta & 0 & -\sin\eta\cos\eta & 0 & \cos 2\eta & 0 \\ 0 & 0 & 0 & -\sin\eta & 0 & \cos\eta \end{bmatrix} \tag{7.6}$$

respectively. The coordinate transformation matrix for a rotation about the crystalline Z axis is the same as the one defined in Eq. (7.4).

7.3 Numerical Results

7.3.1 Cubic inclusions: GaAs/epoxy square lattice

Here, we consider a two-dimensional PC consisting of a square array of circular Gallium Arsenide (GaAs) cylinders embedded in a homogeneous Epoxy background. The PC has a lattice spacing d in both the x and y directions. Bulk acoustic waves propagate along the XY plane [(001) plane] in crystalline GaAs and thus no coordinate transformations need to be applied to the elastic stiffness tensor of the cylinders. GaAs belongs to the cubic crystal system while the Epoxy is isotropic. The elastic stiffness tensor of a cubic material has the form:

$$c = \begin{bmatrix} c_{11} & c_{12} & c_{12} & 0 & 0 & 0 \\ c_{12} & c_{11} & c_{12} & 0 & 0 & 0 \\ c_{12} & c_{12} & c_{11} & 0 & 0 & 0 \\ 0 & 0 & 0 & c_{44} & 0 & 0 \\ 0 & 0 & 0 & 0 & c_{44} & 0 \\ 0 & 0 & 0 & 0 & 0 & c_{44} \end{bmatrix}. \tag{7.7}$$

The slowness surfaces of bulk acoustic waves propagating in the XY plane of GaAs are shown in Fig. 7.2(a). It can be seen that the pure shear mode is isotropic, while the quasi-longitudinal and quasi-shear modes are anisotropic with an eight-fold symmetry. We expect that a cylinder rotation will change the band structure of the PC, including the width and position of the phononic band gaps.

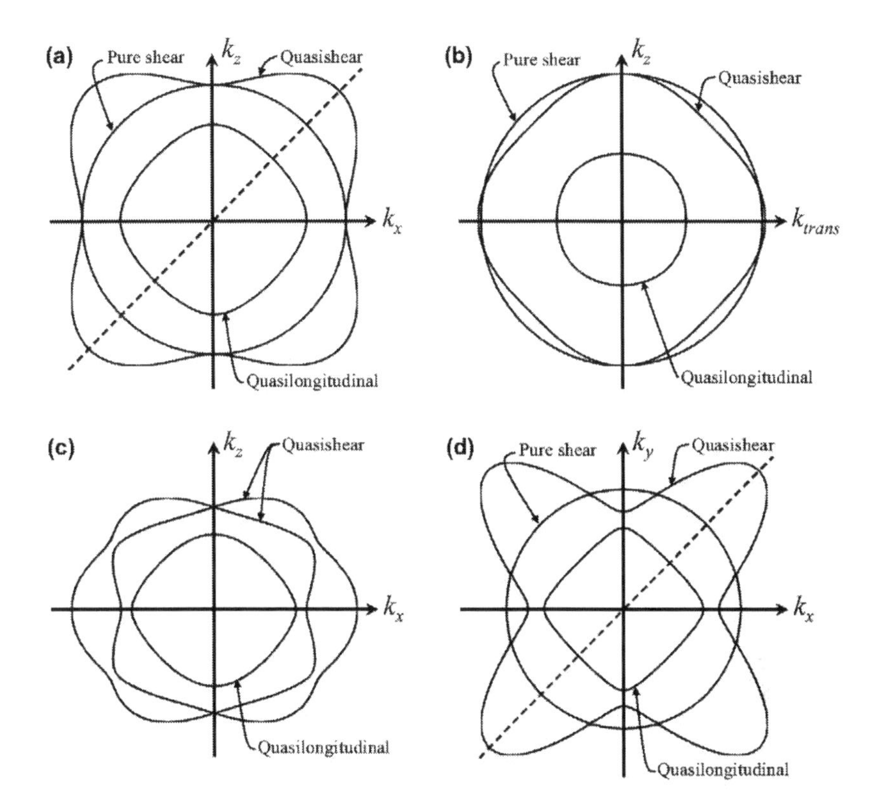

Figure 7.2. Slowness surfaces of bulk acoustic waves propagating in (a) the XZ plane of GaAs, (b) the meridian plane of ZnO, (c) the XZ plane of Quartz, and (d) the XY plane of TiO_2.

Note that the elastic stiffness components $c_{14} = c_{41}$, $c_{15} = c_{51}$, $c_{24} = c_{42}$, $c_{25} = c_{52}$, $c_{46} = c_{64}$, and $c_{56} = c_{65}$ of cubic and isotropic materials are zero; as a result the Fourier components in Eqs. (A.12) and (A.13) vanish ($U^1_{\mathbf{G}-\mathbf{G}'} = U^2_{\mathbf{G}-\mathbf{G}'} = W^1_{\mathbf{G}-\mathbf{G}'} = W^2_{\mathbf{G}-\mathbf{G}'} = 0$). The $\mathbf{M}$ matrix in Eq. (A.8) can therefore be decoupled into two different polarization modes of bulk acoustic waves as

$$\begin{bmatrix} \omega^2 \rho_{\mathbf{G}-\mathbf{G}'} + M^1_{\mathbf{G}-\mathbf{G}'} & L^1_{\mathbf{G}-\mathbf{G}'} \\ L^2_{\mathbf{G}-\mathbf{G}'} & \omega^2 \rho_{\mathbf{G}-\mathbf{G}'} + M^2_{\mathbf{G}-\mathbf{G}'} \end{bmatrix} \cdot \begin{bmatrix} A^1_{\mathbf{G}'} \\ A^2_{\mathbf{G}'} \end{bmatrix} = 0 \tag{7.8}$$

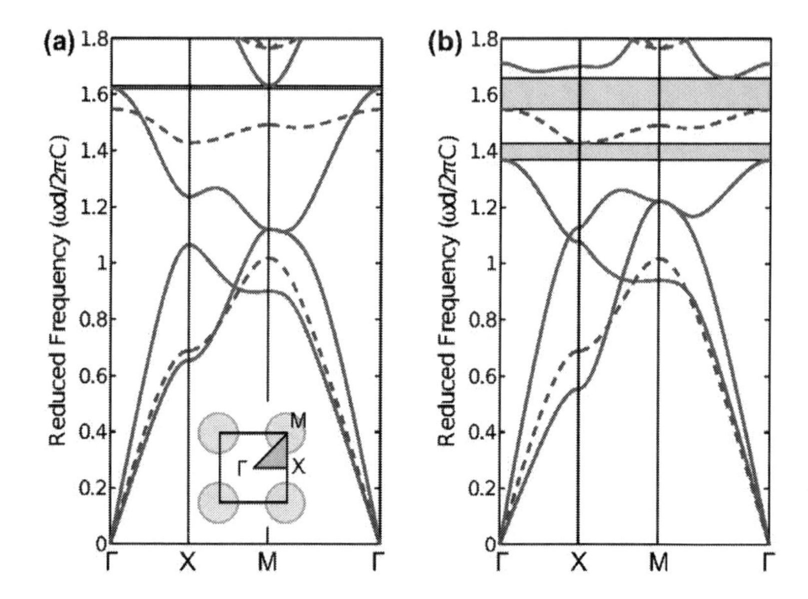

Figure 7.3. Band structure for bulk modes of a GaAs (XZ plane)/epoxy square-lattice PC with a filling fraction of 0.65 and a cylinder rotation angle of (a) 0° and (b) 45°.

for the mixed polarization modes [i.e., longitudinal (L) and shear horizontal (SH) modes with polarization in the xy plane] and

$$\left[\omega^2 \rho_{\mathbf{G}-\mathbf{G}'} + M^3_{\mathbf{G}-\mathbf{G}'}\right] \cdot \left[A^3_{\mathbf{G}'}\right] = 0 \tag{7.9}$$

for the transverse polarization mode [i.e., shear vertical (SV) mode with polarization along the z axis]. The mixed polarization modes correspond to the quasi-longitudinal and quasi-shear modes in GaAs [Fig. 7.2(a)], and the transverse polarization mode corresponds to the pure shear mode.

Figure 7.3(a) shows the calculated band structure for bulk acoustic waves propagating in a GaAs/epoxy PC with a filling fraction of 0.65 and a cylinder rotation angle of $\theta = 0$. The vertical axis of the band structure gives the reduced frequency and the horizontal axis is the reduced wave vector along the periphery of the irreducible part (triangle ΓXM) in the first Brillouin zone. We used 121 reciprocal lattice vectors in all calculations conducted in this article to ensure the convergence of the PWE method. The red solid and blue dashed lines in Fig. 7.3(a) represent the mixed polarization and transverse polarization modes, respectively. The GaAs/epoxy PC demonstrates a narrow phononic band gap ranging from $\Omega = 1.621$ to $\Omega = 1.630$ with a relative bandwidth (phononic band gap width divided by midgap frequency) of 0.55%. The yellow colored area in the figure indicates the location of the phononic band gap. When the GaAs cylinders are rotated through an angle of 45° about the cylinder axis, the elastic stiffness tensor changes with respect to the global coordinate system xyz; the corresponding band structure are displayed in Fig. 7.3(b). Comparing Fig. 7.3(b) with Fig. 7.3(a), the shape of the mixed polarization modes (red solid lines) deforms greatly. However, the band structure for the transverse polarization mode (blue dashed lines) are not altered at all after cylinder rotation because the Fourier components in $M^3_{\mathbf{G}-\mathbf{G}'}$ do not vary with the coordinate transformation. The 45°-cylinder-rotated GaAs/epoxy PC exhibits two phononic band gaps, separated by the second band of the vertical polarization mode, extending from $\Omega = 1.368$ to $\Omega = 1.425$ (relative bandwidth 4.1%) and from $\Omega = 1.548$ to $\Omega = 1.659$ (relative bandwidth 6.9%), respectively.

To further investigate the geometric dependence of the band structure of the GaAs/ epoxy PC, we calculated the midgap frequency and relative bandwidth of the phononic band gaps for cylinder rotation angles from 0 to 45° with an increment of 2.5°, as shown in Figs. 7.4(a) and 7.4(a), respectively. This angular range is sufficient because the angular dependence of the phononic band gaps is symmetric about $\theta = 45°$. We find that the first (lower frequency) phononic band gap only appears when $\theta > 30°$. The midgap frequency of this band gap decreases with θ while the gap width increases with θ. The second (higher frequency) phononic band gap exists throughout all rotation angles; it is located between the third and forth frequency bands of the mixed polarization mode for $\theta \leq 15°$ and between the second band of the transverse polarization mode and the forth frequency band of the mixed polarization mode for $15° < \theta \leq 45°$. The maximum width of the second phononic band gap is 12% at $\theta = 25°$. It is worth noting that the two phononic band gaps are separated by the second band of the transverse polarization mode. The width of the partial phononic band gap of the mixed polarization modes increases progressively with an increase in rotation angle, as displayed in Fig. 7.4(b). The strong tunability of this partial phononic band gap holds great potential for filtering applications.

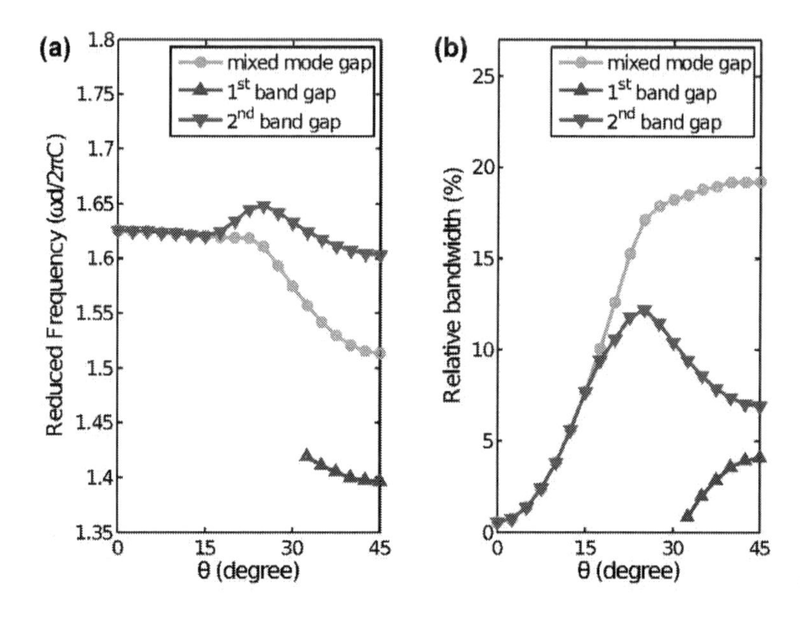

Figure 7.4. (a) and (b) depict the angular dependence of the midgap frequency and relative bandwidth of the phononic band gaps for bulk modes of a GaAs (XZ plane)/epoxy square-lattice PC, respectively.

When compared to existing studies that employ rotation of noncircular cylinders [43, 66] or utilize strong external stimuli [50, 46, 48, 69, 68], our PCs demonstrate a competitive tuning range for both the position and width of the phononic band gaps. Our results suggest that the reorientation of anisotropic cylinders can serve as a feasible approach for tuning the band structure for acoustic waves propagating in PCs.

7.3.2 Hexagonal inclusions: ZnO/epoxy square lattice

Now we consider a two-dimensional, square-lattice PC consisting of Zinc Oxide (ZnO) cylinders embedded in Epoxy. ZnO belongs to the hexagonal crystal system and its elastic stiffness tensor has the form:

$$c = \begin{bmatrix} c_{11} & c_{12} & c_{13} & 0 & 0 & 0 \\ c_{12} & c_{11} & c_{13} & 0 & 0 & 0 \\ c_{13} & c_{13} & c_{33} & 0 & 0 & 0 \\ 0 & 0 & 0 & c_{44} & 0 & 0 \\ 0 & 0 & 0 & 0 & c_{44} & 0 \\ 0 & 0 & 0 & 0 & 0 & (c_{11}-c_{12})/2 \end{bmatrix}. \tag{7.10}$$

Acoustic waves propagate in the meridian plane of crystalline ZnO, thus two coordinate transformations with respect to X and Z axis were applied to the elastic stiffness tensor of the cylinders. Without considering the piezoelectric effect, the slowness surfaces of the three bulk modes at room temperature exhibit a four-fold symmetry shown in Fig. 7.2(b). After coordinate transformations, the elastic stiffness tensor of ZnO is no longer lower than orthorhombic symmetry, hence the **M** matrix in Eq. (A.8) cannot be decoupled into two different polarization modes. The three bulk acoustic modes in the PC are all mixed polarization and distinguished as quasi-L, quasi-SH, and quasi-SV modes.

Figure 7.5(a) shows the calculated band structure for bulk acoustic waves propagating in a ZnO/epoxy PC with a filling fraction of 0.45 and a cylinder rotation angle of $\theta = 0$. Note that the irreducible part of the Brillouin zone for the PC is now rectangular (ΓXMY) due to the four-fold symmetry of the slowness surfaces of ZnO in the meridian plane. It can be seen from Fig.7.5(a) that the band diagrams along Γ-X and X-M are different from those along Γ-Y and Y-M, respectively. The arrows point out differences at X and Y at points near the first phononic band gap. In the figure, we observe that three phononic band gaps exist below reduced frequency $\Omega = 1.5$, between the fourth and fifth, seventh and eighth, and eighth and ninth frequency bands, denoted by the yellow colored areas. The midgap frequency (relative bandwidth) of these three band gaps is 0.7155 (1.82%), 1.2025 (7.07%), and 1.3975 (4.65%), respectively. When the ZnO cylinders are rotated 45° in a clockwise direction, the three band gaps remained between the same pairs of frequency bands [Fig. 7.5(b)]. However, the relative bandwidth of the first and second band gaps expand enormously to 8.01% and 12.98%, respectively, while the third band gap reduces to 2.69%. Figure 7.5 displays the geometric dependence of the phononic band gaps in the ZnO/epoxy PC for a cylinder rotation range of 0 to 90°. The solid lines and shaded areas in Fig. 7.6(a) denote the midgap frequency and the range of the three

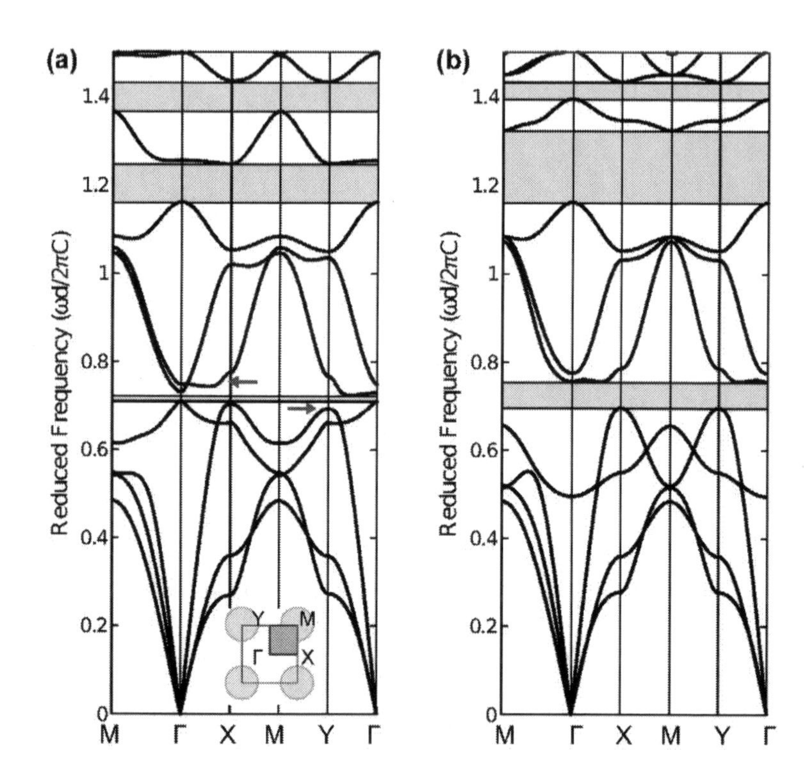

Figure 7.5. Band structure for bulk modes of a ZnO (meridian plane)/epoxy square-lattice PC with a filling fraction of 0.45 and a cylinder rotation angle of (a) 0° and (b) 45°.

phononic band gaps, respectively. The relative bandwidths are plotted in Fig. 7.6(b). At first glance, we notice that the dependence of all the band gaps are symmetric around 45°; the observed symmetry can be explained by the fact that our calculations account for wave propagation in the entire ΓXMY rectangle of the first Brillouin zone. For example, the band diagrams along Γ-X at θ=0 are identical to those along Γ-Y at $\theta = 90°$ due to the four-fold symmetry of the slowness surfaces of ZnO. The width of the first phononic band gap increases monotonically with cylinder rotation angle and reaches a maximum of 8.01% at $\theta = 45°$. The widths of the second and third band gaps also vary with rotation angle,

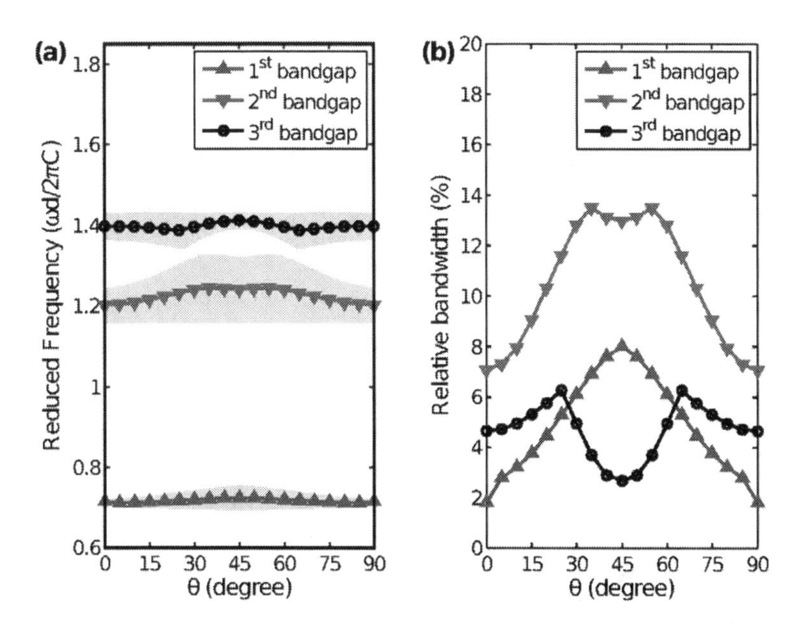

Figure 7.6. Angular dependence of (a) the midgap frequency and (b) relative bandwidth of phononic band gaps of a ZnO (meridian plane)/epoxy square-lattice PC with a filling fraction of 0.45.

but their maxima appear at $\theta = 35°$ (and $\theta = 55°$ due to the mirror-image symmetry) and $\theta = 25°$ ($\theta = 65°$), respectively. Figure 7.6 demonstrates that the proposed method can be used to tune the phononic band gaps of a square-lattice PC consisting of hexagonal cylinders over a considerable range.

7.3.3 Trigonal inclusions: Quartz/epoxy square lattice

In this subsection, we investigate a PC consisting of a two-dimensional square array of Quartz cylinders embedded in a homogeneous Epoxy background. Quartz belongs to the trigonal crystal system, which has six independent elastic stiffness components (see Table 7.1). Bulk acoustic waves propagate in the XZ plane [(010) plane] of Quartz, thus a coordinate transformation with respect to the crystalline X axis is applied to the elastic

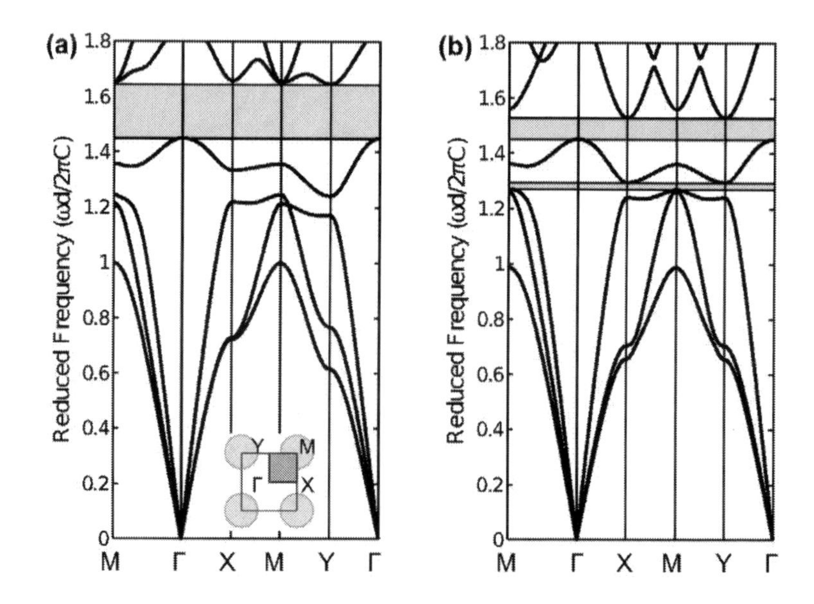

Figure 7.7. Band structure for bulk modes of a Quartz (XZ plane)/epoxy square-lattice PC with a filling fraction of 0.6 and a cylinder rotation angle of (a) 90° and (b) 45°.

stiffness tensor of the cylinders. The slowness surfaces of the bulk modes propagating in the XZ plane of Quartz have a four-fold symmetry, as shown in Fig.7.2(c). The band structure of a Quartz/epoxy PC with a filling fraction of 0.6 and a cylinder rotation angle of 90° are shown in Fig. 7.7(a). In the figure, we observe a wide phononic band gap between the fourth and fifth frequency bands that expands in reduced frequency from 1.448 to 1.644. The midgap frequency and relative bandwidth of the band gap are 1.546 and 19.6%, respectively. Note that if only the ΓXM triangle in the first Brillouin zone is considered, there should be a phononic band gap between the third and fourth frequency bands; however, the overlap of these frequency bands in the Y-M direction closes this gap. Figure 7.7(b) shows the band structure when the cylinder rotation angle is 45°. At this cylinder rotation angle the third and fourth frequency bands do not overlap and a phononic band gap exists, extending from 1.270 to 1.294 with a relative bandwidth of 2.4%. The second phononic band gap remains intact between the fourth and fifth

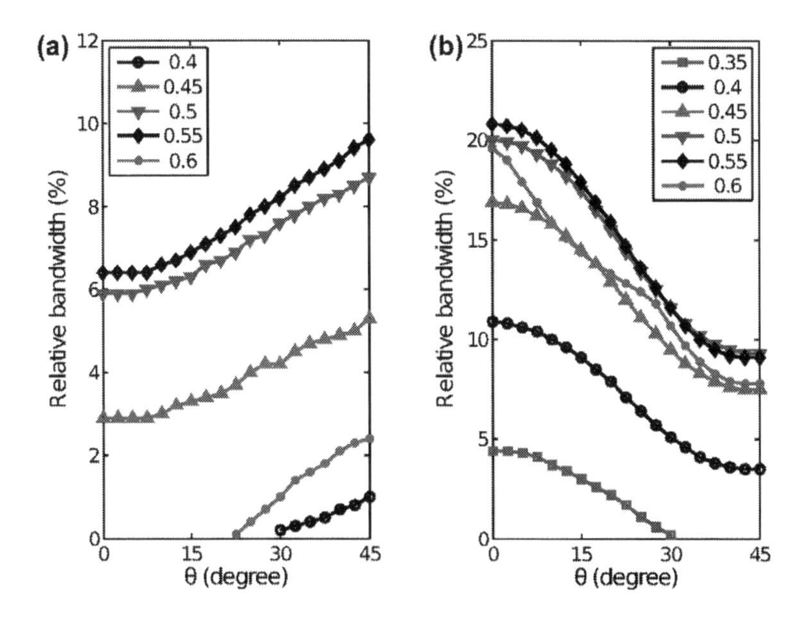

Figure 7.8. Angular dependence of the relative bandwidth of (a) first and (b) second phononic band gaps of Quartz (XZ plane)/epoxy square-lattice PCs with different filling fractions.

frequency bands, however the relative bandwidth of the phononic band gap decreases to 7.8%.

In periodic composite structures, the filling fraction can be another important parameter on band structure deformation and band gap modification. Figure 7.8 shows the relative bandwidth of the first [Fig. 7.8(a)] and second [Fig. 7.8(b)] phononic band gaps in the Quartz/epoxy PCs as a function of filling fraction and cylinder rotation angle. Only filling fractions ranging from 0.35 to 0.6 (with an increment of 0.05) are shown because no phononic band gap exists in Quartz/epoxy PCs for filling fractions outside this range. For any fixed θ, we can see that the width of both phononic band gaps increases with filling fraction, reaching a maximum at $f = 0.55$ (black lines with diamond markers), then decreasing and finally disappearing when $f > 0.6$. In this regard, a Quartz/epoxy PC can be designed with a filling fraction close to 0.55 to obtain the widest possible

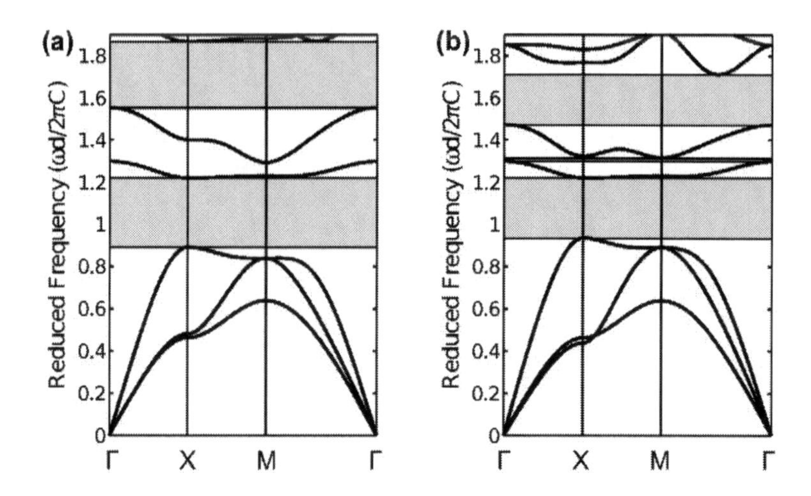

Figure 7.9. Band structure for bulk modes of a TiO_2 (XY plane)/epoxy square-lattice PC with a filling fraction of 0.5 and a cylinder rotation angle of (a) 0° and (b) 45°.

phononic band gap. For all filling fractions shown in the figure, we observe that the first phononic band gap width climbs with cylinder rotation angle [Fig. 7.8(a)], while the second phononic band gap width declines with cylinder rotation angle [Fig. 7.8(b)]. In fact, the first band gap of the PC with $f = 0.4$ and 0.6 does not open until θ increases to 22.5° and 30°, respectively. The second band gap of the PC with $f = 0.35$ closes when $\theta > 30°$. Therefore, by choosing a proper filling fraction and controlling the rotation angle of the Quartz cylinders, one can obtain a tunable PC for filtering or waveguiding applications.

7.3.4 Tetragonal inclusions: TiO_2/epoxy square lattice

Finally in this section we consider a two-dimensional, square-lattice PC consisting of Rutile (TiO_2) cylinders embedded in Epoxy. The TiO_2 belongs to the tetragonal crystal system that has six independent elastic stiffness components. Bulk acoustic waves propagate in the XY plane of crystalline TiO_2, hence no coordinate transformation is necessary. The slowness surfaces of the bulk modes propagating in TiO_2 are shown in Fig. 7.2(d); they exhibit an eight-fold symmetry. The band structure of a TiO_2/epoxy PC with a fill-

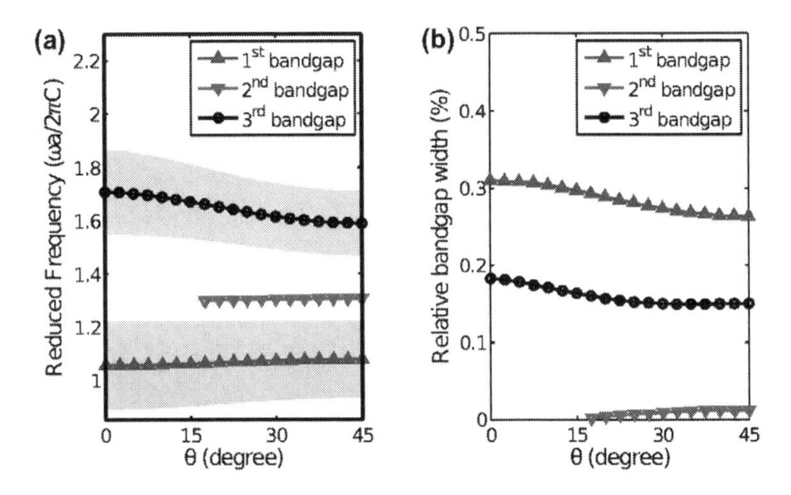

Figure 7.10. (a) and (b) depict the angular dependence of the midgap frequency and relative bandwidth of the phononic band gaps for bulk modes of a TiO_2 (XY plane)/epoxy square-lattice PC, respectively.

ing fraction of 0.5 and cylinder rotation angles of 0° and 45° are shown in Figs. 7.9(a) and 7.9(b). Figures 7.10(a) and 7.10(b) display the angular dependence of position and width of the band gaps respectively. From these figures, we observe that two wide phononic band gaps exist for all cylinder rotation angles and a very narrow phononic band gap between the fourth and fifth frequency bands opens up at $\theta > 17.5°$. These calculations demonstrate the tunability of PCs composed of cylinders with high anisotropy.

7.4 Summary

In this chapter, we present a comprehensive study of tunable phononic band gaps in two-dimensional square-lattice PCs consisting of anisotropic inclusions in an isotropic host. The anisotropic materials considered in this study include cubic, hexagonal, trigonal, and tetragonal crystal systems. We have numerically observed that the band structure for bulk acoustic waves propagating in such heterogeneous structures were greatly deformed upon rotation of the anisotropic cylinders. The tunable range of the band gap width and location demonstrated in this study are competitive with methods that utilize the rotation of noncircular cylinders and much greater than methods using external stimuli. From our theoretical investigations, we suggest that the reorientation of anisotropic cylinders can be a simple and effective way to obtain selective filtering and waveguiding for acoustic metamaterial applications such as acoustic imaging, high intensity focused ultrasound, and nondestructive evaluation (NDE).

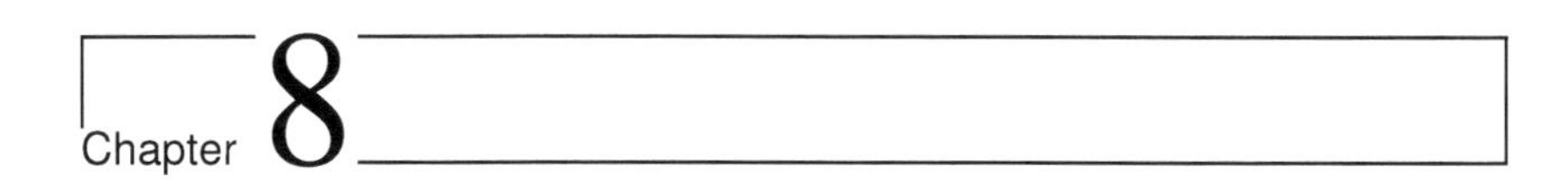

Conclusions and Prospects

8.1 Achievements

Artificially engineered periodic elastic structures, or phononic crystals (PCs), offers an opportunity to manipulate the propagation of acoustic waves in a way not be found in nature. Recently, there have been advances in developing PC-based acoustic mirrors, high-efficiency waveguides, frequency-selective filters, and subwavelength lenses. However, before we are able to apply these techniques to real applications, more functionalities need to be demonstrated and some fabrication considerations shall be fulfilled. For example, one of the principle limitations to applied current PC techniques to audible noise control is the size mismatch problem. Though PCs can perfectly reflect sound within the phononic band gaps, the thickness of these PCs are on the scale of centimeters to meters which makes it impractical to build PC-based headphones or earplugs. The narrow working frequency range of PC-based acoustic components also hinders the extension of these promising acoustic metamaterials to our daily life applications.

This book has centered on inventing a new class of PCs to benefit a wide range of applications from communications to medical diagnosis and therapy. By introducing the "gradient-index" concept to regular PCs, I have successfully demonstrated a number of gradient-index phononic crystals (GRIN PCs) that could achieve unprecedented wave manipulation behaviors over a wide working band. The footprint of these GRIN PCs are much smaller than their PC counterparts, therefore the techniques shown in this book has taken a big step toward the realization of PC-based acoustic devices for sound control. Further, a practical tuning method has been proposed to transform passive PC-based acoustic components into active acoustic devices.

Taking a systematic approach, several new functionalities have been demonstrated by GRIN PCs. A quick review of my research achievements presented in this book is: wide-band acoustic self-collimation by a PC composite, wavelength-scale acoustic mirage inside a GRIN PC with linearly modulated filling fraction, subwavelength wide-band focusing by a flat GRIN PC lens with a hyperbolic-secant gradient profile, GRIN PC-based high-coupling-rate acoustic beamwidth compressor, high-efficiency acoustic beam aperture modifier by a butt-jointed GRIN PC composite, and tunable PCs with anisotropic inclusions. These research results have been published in high-profile journals such as *Physical Review B*, *Applied Physics Letters*, *Journal of Applied Physics*, and *Journal of Physics D: Applied Physics* [17, 55, 59, 51, 60, 65, 64], and, due to their significance, were featured as the front cover images of the journal, as shown in Fig 8.1.

8.2 Prospects

Despite recent successful developments in acoustic metamaterials, the field remains in its infancy and not a killer application has been demonstrated. Lots of works need to be done before we can finally enjoy the advance of PC technology in our everyday lives. Firstly, the GRIN PC concepts presented in this book are numerically verified mainly for SV-mode BAW only. Though the GRIN concept is universal, as it has also been applied to

Figure 8.1. Our works are featured as the front cover image of *Applied Physics Letters*, *Journal of Physics D: Applied Physics*, and *Journal of Applied Physics*.

photonic crystals for different electromagnetic wave modes and surface plasmons [54, 71, 72], GRIN PCs for longitudinal waves, surface waves, and Lamb waves need be designed and experimentally evaluated.

Secondly, the tunable PC concept proposed in this book can be combined with the GRIN PC concept to produce a line of tunable GRIN PCs whose acoustic beam focusing or bending behavior can be dynamically tuned when operating at a fixed working frequency (e.g., at design frequency for best performance).

Thirdly, the GRIN concept is currently applied to regular PCs consist of two different materials. Even the overall size of a GRIN PC has been decreased several times smaller from it's PC counterparts, GRIN PCs are still too big to be considered as practical materials for noise control. To address the size mismatch problem, we will apply the GRIN concept to another class of acoustic metamaterials—locally resonant sonic materials [73]—to further decrease the footprint of an acoustic metamaterial.

Finally, phononic band gaps and photonic band gaps have been observed to co-exist in a PC working at ultra-high frequency ranges. As a result, they are named "phoxonic crystals," the combination of phononic crystals and photonic crystals. By applying the GRIN concept to phoxonic crystals, one may be able to control the acoustic propagation by optical waves, or vise versa. Along the line, new science may be found and powerful devices could be development.

Appendix A

Two-Dimensional Plane Wave Expansion Method

A.1 Formation

In an inhomogeneous linear elastic anisotropic medium with no body force and temperature factor, the equation of motion for the displacement vector $\mathbf{u}(\mathbf{r}, t)$ can be written as

$$\rho(\mathbf{r})\ddot{u}_i(\mathbf{r}, t) = \partial_j[c_{ijmn}(\mathbf{r})\partial_n u_m(\mathbf{r}, t)], \tag{A.1}$$

where $\mathbf{r} = (\mathbf{x}, z) = (x, y, z)$ is the position vector, t is the time variable, $\rho(\mathbf{r})$ and $c_{ijmn}(\mathbf{r})$ are the position-dependent mass density and elastic stiffness tensor, respectively.

Consider a two-dimensional PC composed of a periodic array (x-y plane) of material A embedded in a background material B. Both materials A and B are crystals with the lowest symmetry, i.e., belonging to the triclinic symmetry. Due to the spatial periodicity, the material constants, $\rho(\mathbf{x})$ and $c_{ijmn}(\mathbf{x})$ can be expanded in the Fourier series with respect to the two-dimensional reciprocal lattice vectors, $\mathbf{G} = (G_1, G_2)$, as

$$\rho(\mathbf{x}) = \sum_{\mathbf{G}} e^{i\mathbf{G}\cdot\mathbf{x}} \rho_{\mathbf{G}}, \tag{A.2}$$

$$C_{ijmn}(\mathbf{x}) = \sum_{\mathbf{G}} e^{i\mathbf{G}\cdot\mathbf{x}} C_{\mathbf{G}}^{ijmn}, \tag{A.3}$$

where $\rho_{\mathbf{G}}$ and $C_{\mathbf{G}}^{ijmn}$ are the corresponding Fourier coefficients and are defined as

$$\rho_{\mathbf{G}} = A_c^{-1} \int d^2x \rho(\mathbf{x}) e^{-i\mathbf{G}\cdot\mathbf{x}}, \tag{A.4}$$

$$C_{\mathbf{G}}^{ijmn} = A_c^{-1} \int d^2x C_{ijmn}(\mathbf{x}) e^{-i\mathbf{G}\cdot\mathbf{x}}, \tag{A.5}$$

In the above equations, A_c is the area of the unit cell of the PC.

On expanding the displacement vector and utilizing the Bloch theorem,

$$\mathbf{u}(\mathbf{r}, t) = \sum_{\mathbf{G}} e^{i\mathbf{k}\cdot\mathbf{x} - i\omega t}(e^{i\mathbf{G}\cdot\mathbf{x}} \mathbf{A}_{\mathbf{G}} e^{ik_z z}), \tag{A.6}$$

where $\mathbf{k} = (k_1, k_2)$ is the Block wave vector, ω is the angular frequency, k_z is the wave number along the z axis, and $\mathbf{A}_{\mathbf{G}}$ is the amplitude of the displacement vector. As the component of the wave vector k_z equals to zero, Eq. (A.6) degrades to the displacement vector of a bulk acoustic wave.

Substituting Eqs. (A.2), (A.3), and (A.6) into Eq. (A.1) and consider bulk acoustic modes only, we obtain an eigenvalue problem [7, 11]:

$$\mathbf{MU} = 0, \tag{A.7}$$

$$\mathbf{M} = \begin{bmatrix} \omega^2 \rho_{\mathbf{G}-\mathbf{G}'} + M^1_{\mathbf{G}-\mathbf{G}'} & L^1_{\mathbf{G}-\mathbf{G}'} & U^1_{\mathbf{G}-\mathbf{G}'} \\ L^2_{\mathbf{G}-\mathbf{G}'} & \omega^2 \rho_{\mathbf{G}-\mathbf{G}'} + M^2_{\mathbf{G}-\mathbf{G}'} & U^2_{\mathbf{G}-\mathbf{G}'} \\ W^1_{\mathbf{G}-\mathbf{G}'} & W^2_{\mathbf{G}-\mathbf{G}'} & \omega^2 \rho_{\mathbf{G}-\mathbf{G}'} + M^3_{\mathbf{G}-\mathbf{G}'} \end{bmatrix}, \tag{A.8}$$

$$\mathbf{U} = \begin{bmatrix} A^1_{\mathbf{G}'} \\ A^2_{\mathbf{G}'} \\ A^3_{\mathbf{G}'} \end{bmatrix}, \tag{A.9}$$

where $\mathbf{U}$ is the eigenvector and $\mathbf{M}$ is a function of the Block wave vector $\mathbf{k}$, angular frequency ω, Fourier coefficients of mass density $\rho_{\mathbf{G}-\mathbf{G}'}$ and components of elastic stiffness tensor $c^{ij}_{\mathbf{G}-\mathbf{G}'}$. The expressions of the nine matrix entries in Eq. (A.8) are

$$\begin{aligned} M^1_{\mathbf{G}-\mathbf{G}'} &= [b_1 c^{11}_{\mathbf{G}-\mathbf{G}'} + b_2 c^{16}_{\mathbf{G}-\mathbf{G}'} + b_3 c^{61}_{\mathbf{G}-\mathbf{G}'} + b_4 c^{66}_{\mathbf{G}-\mathbf{G}'}], \\ M^2_{\mathbf{G}-\mathbf{G}'} &= [b_1 c^{66}_{\mathbf{G}-\mathbf{G}'} + b_2 c^{62}_{\mathbf{G}-\mathbf{G}'} + b_3 c^{26}_{\mathbf{G}-\mathbf{G}'} + b_4 c^{22}_{\mathbf{G}-\mathbf{G}'}], \\ M^3_{\mathbf{G}-\mathbf{G}'} &= [b_1 c^{55}_{\mathbf{G}-\mathbf{G}'} + b_2 c^{54}_{\mathbf{G}-\mathbf{G}'} + b_3 c^{45}_{\mathbf{G}-\mathbf{G}'} + b_4 c^{44}_{\mathbf{G}-\mathbf{G}'}], \end{aligned} \tag{A.10}$$

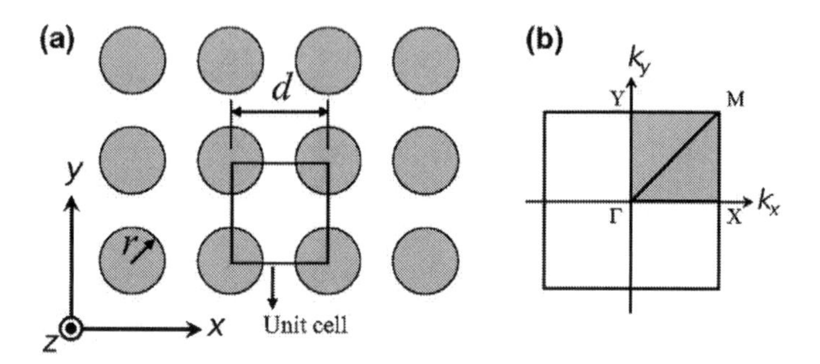

Figure A.1. (a) Top view of an infinite two-dimensional square-lattice PC. (b) The first Brillouin zone of the unit cell.

$$
\begin{aligned}
L^1_{\mathbf{G}-\mathbf{G}'} &= [b_1 c^{16}_{\mathbf{G}-\mathbf{G}'} + b_2 c^{12}_{\mathbf{G}-\mathbf{G}'} + b_3 c^{66}_{\mathbf{G}-\mathbf{G}'} + b_4 c^{62}_{\mathbf{G}-\mathbf{G}'}], \\
L^2_{\mathbf{G}-\mathbf{G}'} &= [b_1 c^{16}_{\mathbf{G}-\mathbf{G}'} + b_2 c^{66}_{\mathbf{G}-\mathbf{G}'} + b_3 c^{21}_{\mathbf{G}-\mathbf{G}'} + b_4 c^{26}_{\mathbf{G}-\mathbf{G}'}],
\end{aligned} \tag{A.11}
$$

$$
\begin{aligned}
U^1_{\mathbf{G}-\mathbf{G}'} &= [b_1 c^{15}_{\mathbf{G}-\mathbf{G}'} + b_2 c^{14}_{\mathbf{G}-\mathbf{G}'} + b_3 c^{65}_{\mathbf{G}-\mathbf{G}'} + b_4 c^{64}_{\mathbf{G}-\mathbf{G}'}], \\
U^2_{\mathbf{G}-\mathbf{G}'} &= [b_1 c^{65}_{\mathbf{G}-\mathbf{G}'} + b_2 c^{64}_{\mathbf{G}-\mathbf{G}'} + b_3 c^{25}_{\mathbf{G}-\mathbf{G}'} + b_4 c^{24}_{\mathbf{G}-\mathbf{G}'}],
\end{aligned} \tag{A.12}
$$

$$
\begin{aligned}
W^1_{\mathbf{G}-\mathbf{G}'} &= [b_1 c^{51}_{\mathbf{G}-\mathbf{G}'} + b_2 c^{56}_{\mathbf{G}-\mathbf{G}'} + b_3 c^{41}_{\mathbf{G}-\mathbf{G}'} + b_4 c^{46}_{\mathbf{G}-\mathbf{G}'}], \\
W^2_{\mathbf{G}-\mathbf{G}'} &= [b_1 c^{56}_{\mathbf{G}-\mathbf{G}'} + b_2 c^{52}_{\mathbf{G}-\mathbf{G}'} + b_3 c^{46}_{\mathbf{G}-\mathbf{G}'} + b_4 c^{42}_{\mathbf{G}-\mathbf{G}'}],
\end{aligned} \tag{A.13}
$$

where

$$
\begin{aligned}
b_1 &= -(G_1 + k_1)(G'_1 + k_1), \quad b_2 = -(G_1 + k_1)(G'_2 + k_2), \\
b_3 &= -(G_2 + k_2)(G'_1 + k_1), \quad b_4 = -(G_2 + k_2)(G'_2 + k_2).
\end{aligned} \tag{A.14}
$$

In the above equations, Voigt's notation has been used to express the Fourier coefficients of the elastic stiffness tensor $c^{ij}_{\mathbf{G}-\mathbf{G}'}$. The eigenfrequencies of the bulk acoustic modes can be obtained by setting

$$
det(\mathbf{M}) = 0. \tag{A.15}
$$

Once the eigenfrequencies are obtained, the relative amplitude of the displacement for each eigenmode can be solved by substituting the eigenfrequencies into Eq. (A.7).

A.2 Structure Factors

The Fourier coefficients $\rho_{\mathbf{G}}$ and $C_{\mathbf{G}}^{ijmn}$ in Eqs. (A.2) to (A.5) of a PC can be expressed as

$$\alpha_{\mathbf{G}} = \begin{cases} \alpha_A f + \alpha_B(1-f) & \text{for } \mathbf{G} = 0 \\ (\alpha_A - \alpha_B) S_{\mathbf{G}} & \text{for } \mathbf{G} \neq 0 \end{cases} \tag{A.16}$$

where $\alpha_{\mathbf{G}} = (\rho_{\mathbf{G}}, C_{\mathbf{G}}^{ijmn})$, f is the filling fraction of the cylinder, and $S_{\mathbf{G}}$ is the structure function.

For a two-dimensional square-lattice PC with circular cylinders (Fig. A.1), the filling fraction $f = \pi r^2/a^2$ and the area of the unit cell $A_c = d^2$, where d is the lattice spacing. The structure function is

$$S_{\mathbf{G}} = \frac{2f J_1(Gr)}{Gr}, \tag{A.17}$$

where $J_1(x)$ is a first order Bessel function and r is the radius of cylinders.

A.3 Convergence

The accuracy of the PWE method depends on the number of the terms used in the series expansion, i.e., the number of plane waves (reciprocal lattice vectors) used in the band structure calculation. In general, the more plane waves used, the more accurate in the results. However, employing a large number of plane waves in the calculation leads to great amount of time consumed in computation. A balance between satisfactorily convergent results and computational loads should be made. Based on our experience, 121 plane waves can ensure the convergence of the PWE method for solid-solid square-lattice PCs.

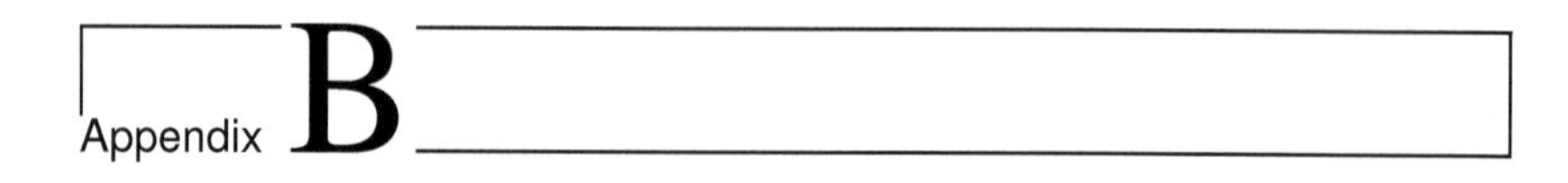

Finite Difference Time Domain Method

B.1 Wave Propagation Computation

The FDTD method is widely used to calculate the transient wave propagation of acoustic and electromagnetic waves, and is suitable for dealing inhomogeneous structures with complex boundaries, such as PCs. The FDTD program was developed using the theory of elasticity: the equation of motion

$$\rho \ddot{u}_i = \tau_{ij,j} + \rho f_i \tag{B.1}$$

and the constitutive law

$$\tau_{ij} = C_{ijkl} \epsilon_{kl} \tag{B.2}$$

are discretized to simulate wave propagation in linear elastic materials, where ρ, τ_{ij}, f_i, C_{ijkl}, and ϵ_{kl} are density of materials, stress, body force, elastic constant, and strain, respectively. The strain can be described based on small deformation with the restrictions of linear theory, and the strain-displacement relation is

$$\epsilon_{kl} = \frac{1}{2} \left(u_{i,j} + u_{j,i} \right). \tag{B.3}$$

The elastodynamic equations are first expanded into Taylor's series to develop second-order difference approximations, and then converted to first-order difference equations. Discrete displacement and stress components are placed in staggered grids (Fig. B.1) and computed recursively with time to study the wave propagation in the simulation domain.

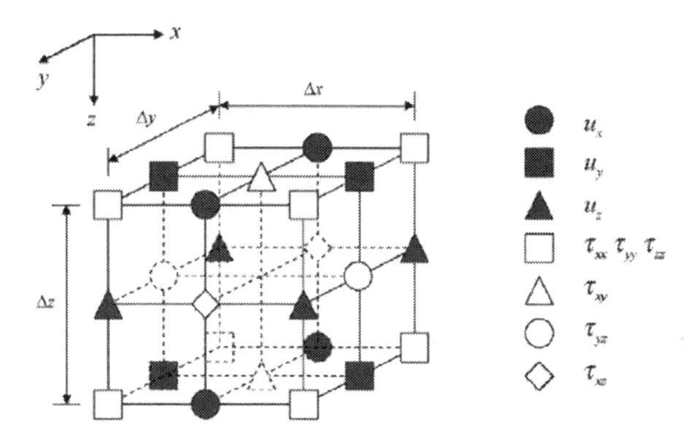

Figure B.1. The unit of the staggered grids in the FDTD method. (Source of image: doi: 10.1109/TUFFC.2006.1588400)

To use FDTD, a simulation domain must be established and discretized into grids. Each grid unit must be specified by defining the density and elastic constants of the material. Finer grids (higher number of total grids) produce more accurate simulation results, but consume longer time to finish the computation. In general, a discrete grid size of 1/10–1/20 of the wavelength is at balance.

A wave source can be defined by setting the initial body force [the last term in the right-hand side of Eq. (B.1)] in the equation of motion. To excite SV-mode BAW, for example, an initial value of body force in the equation of motion along the z axis (out-of-plane direction) is defined. In study of infinite space cases, absorbing boundary conditions were developed to avoid the reflection of waves at simulation domain boundaries. In practice, perfectly matched layers (PML) are set around the simulation domain with damping properties to create non-reflection boundaries. A typical PML thickness used in this book is 20–40 layers, and the reflection coefficient of max amplitude can be reduced to 1.5% or less [24]

B.2 Band Structure Calculation

The FDTD method has been adapted to calculate the band structure of a PC by defining periodic boundary condition of a unit cell of the PC using the Block's theorem. Therefore, only a small computation domain (one unit cell) is needed in the calculation. The phase difference between periodic boundaries is calculated by the wave vector. For a given wave vector, a small disturbance at a random location in the unit cell is set to excite a wide-band signal at the initial time step. The displacement fields are recorded at several locations inside the unit cell and then transformed into frequency domain to find the frequency spectra. The resonance peaks of the frequency spectra are the eigenfrequencies of the PC for the given wave vector. The wave vector must be varied in order to obtain a complete map of band structure.

The geometry and material properties of a PC can be easily assigned in the FDTD method. The computation time for solving the band structure of three-dimensional PCs or complex two-dimensional PCs is usually less than the PWE method. Hence the FDTD method is often used to analyze complicate PC structures such as PC waveguides and PC cavities with the concept of super-cell.

B.3 Acknowledgement

The FDTD program used to simulate the wave propagation in PCs and to calculate the band structure of PCs (in Chapter 5 and 6) in this book was developed by Dr. Jia-Hong Sun (Professor of Mechanical Engineering at Chang Gung University) and Dr. Tsung-Tsong Wu (Professor of Applied Mechanics at National Taiwan University). We gratefully acknowledge the grant of use permission.

Bibliography

[1] SIGALAS, M. and E. N. ECONOMOU (1993) "Band structure of elastic waves in two dimensional systems," *Solid State Communications*, **86**(3), pp. 141–143.

[2] KUSHWAHA, M. S., P. HALEVI, L. DOBRZYNSKI, and B. DJAFARI-ROUHANI (1993) "Acoustic band structure of periodic elastic composites," *Physical Review Letters*, **71**(13), pp. 2022–2025.

[3] VINES, R. E., J. P. WOLFE, and A. V. EVERY (1999) "Scanning phononic lattices with ultrasound," *Physical Review B*, **60**(17), pp. 11871–11874.

[4] PSAROBAS, I. E., N. STEFANOU, and A. MODINOS (2000) "Scattering of elastic waves by periodic arrays of spherical bodies," *Physical Review B*, **62**(1), pp. 278–291.

[5] CERVERA, F., L. SANCHIS, J. V. SÁNCHEZ-PÉREZ, R. MARTÍNEZ-SALA, C. RUBIO, F. MESEGUER, C. LÓPEZ, D. CABALLERO, and J. SÁNCHEZ-DEHESA (2001) "Refractive Acoustic Devices for Airborne Sound," *Physical Review Letters*, **88**(2), p. 023902.

[6] RUZZENE, M., F. SCARPA, and F. SORANNA (2003) "Wave beaming effects in two-dimensional cellular structures," *Smart Materials and Structures*, **12**(3), p. 363.

[7] WU, T.-T., Z.-G. HUANG, and S.-C. S. LIN (2004) "Surface and bulk acoustic waves in two-dimensional phononic crystal consisting of materials with general anisotropy," *Physical Review B*, **69**(9), p. 094301.

[8] WANG, G., J. WEN, Y. LIU, and X. WEN (2004) "Lumped-mass method for the study of band structure in two-dimensional phononic crystals," *Physical Review B*, **69**(18), p. 184302.

[9] LAUDE, V., M. WILM, S. BENCHABANE, and A. KHELIF (2005) "Full band gap for surface acoustic waves in a piezoelectric phononic crystal," *Physical Review E*, **71**(3), p. 036607.

[10] WU, T.-T., Z.-C. HSU, and Z.-G. HUANG (2005) "Band gaps and the electromechanical coupling coefficient of a surface acoustic wave in a two-dimensional piezoelectric phononic crystal," *Physical Review B*, **71**(6), p. 064303.

[11] HSU, J.-C. and T.-T. WU (2006) "Efficient formulation for band-structure calculations of two-dimensional phononic-crystal plates," *Physical Review B*, **74**(14), p. 144303.

[12] YAN, Z.-Z. and Y.-S. WANG (2006) "Wavelet-based method for calculating elastic band gaps of two-dimensional phononic crystals," *Physical Review B*, **74**(22), p. 224303.

[13] SUTTER-WIDMER, D., S. DELOUDI, and W. STEURER (2007) "Prediction of Bragg-scattering-induced band gaps in phononic quasicrystals," *Physical Review B*, **75**(9), p. 094304.

[14] HOU, Z. and B. M. ASSOUARA (2008) "Modeling of Lamb wave propagation in plate with two-dimensional phononic crystal layer coated on uniform substrate using plane-wave-expansion method," *Physics Letters A*, **372**(12), pp. 2091–2097.

[15] SAINIDOU, R., B. DJAFARI-ROUHANI, and J. O. VASSEUR (2008) "Surface acoustic waves in finite slabs of three-dimensional phononic crystals," *Physical Review B*, **77**(9), p. 094304.

[16] GAO, J., X.-Y. ZOU, J.-C. CHENG, and B. LI (2008) "Band gaps of lower-order Lamb wave in thin plate with one-dimensional phononic crystal layer: Effect of substrate," *Applied Physics Letters*, **92**(2), p. 023510.

[17] SHI, J., S.-C. S. LIN, and T. J. HUANG (2008) "Wide-Band Acoustic Collimating by Phononic Crystal Composites," *Applied Physics Letters*, **92**(11), p. 111901.

[18] YANG, S., J. H. PAGE, Z. LIU, M. L. COWAN, C. T. CHAN, and P. SHENG (2004) "Focusing of Sound in a 3D Phononic Crystal," *Physical Review Letters*, **93**(2), p. 024301.

[19] KE, M., Z. LIU, C. QIU, W. WANG, J. SHI, W. WEN, and P. SHENG (2005) "Negative-refraction imaging with two-dimensional phononic crystals," *Physical Review B*, **72**(6), p. 064306.

[20] LI, J., Z. LIU, and C. QIU (2006) "Negative refraction imaging of acoustic waves by a two-dimensional three-component phononic crystal," *Physical Review B*, **73**(5), p. 054302.

[21] CAI, L.-W., D. K. DACOL, D. C. CALVO, and G. J. ORRIS (2007) "Acoustical scattering by arrays of cylinders in waveguides," *Journal of Acoustical Society of America*, **122**(3), pp. 1340–1351.

[22] MARTIN, T. P., M. NICHOLAS, G. J. ORRIS, L.-W. CAI, D. TORRENT, and J. SÁNCHEZ-DEHESA (2010) "Sonic gradient index lens for aqueous applications," *Applied Physics Letters*, **97**(11), p. 113503.

[23] SUN, J.-H. and T.-T. WU (2005) "Analyses of mode coupling in joined parallel phononic crystal waveguides," *Physical Review B*, **71**(17), p. 174303.

[24] HSIEH, P.-F., T.-T. WU, and J.-H. SUN (2006) "Three-dimensional phononic band gap calculations using the FDTD method and a PC cluster system," *ieee transactions on ultrasonics, ferroelectrics, and frequency control*, **53**(1), pp. 148–158.

[25] SUN, J.-H. and T.-T. WU (2006) "Propagation of surface acoustic waves through sharply bent two-dimensional phononic crystal waveguides using a finite-difference time-domain method," *Physical Review B*, **74**(17), p. 174305.

[26] TANAKA, Y., Y. TOMOYASU, and S. ICHIRO TAMURA (2000) "Band structure of acoustic waves in phononic lattices: Two-dimensional composites with large acoustic mismatch," *Physical Review B*, **62**(11), pp. 7387–7392.

[27] SIGALAS, M. M. (1998) "Defect states of acoustic waves in a two-dimensional lattice of solid cylinders," *Journal of Applied Physics*, **84**(6), p. 3026.

[28] BRIA, D. and B. DJAFARI-ROUHANI (2002) "Omnidirectional elastic band gap in finite lamellar structures Omnidirectional elastic band gap in finite lamellar structures," *Physical Review E*, **66**(5), p. 056609.

[29] KHELIF, A., B. DJAFARI-ROUHANI, J. O. VASSEUR, and P. A. DEYMIER (2003) "Transmission and dispersion relations of perfect and defect-containing waveguide structures in phononic band gap materials," *Physical Review B*, **68**(2), p. 024302.

[30] KHELIF, A., A. CHOUJAA, S. BENCHABANE, B. DJAFARI-ROUHANI, and V. LAUDE (2004) "Guiding and bending of acoustic waves in highly confined phononic crystal waveguides," *Applied Physics Letters*, **84**(22), pp. 4400–4402.

[31] YAO, Y., Z. HOU, and Y. LIU (2006) "The propagating properties of the heterostructure phononic waveguide," *Journal of Physics D: Applied Physics*, **39**(24), p. 5164.

[32] LIN, K.-H., C.-F. CHANG, C.-C. PAN, J.-I. CHYI, S. KELLER, U. MISHRA, S. P. DENBAARS, and C.-K. SUN (2006) "Characterizing the nanoacoustic superlattice in a phonon cavity using a piezoelectric single quantum well," *Applied Physics Letters*, **89**(14), p. 143103.

[33] TANAKA, Y., T. YANO, and S. ICHIRO TAMURA (2007) "Surface guided waves in two-dimensional phononic crystals," *Wave Motion*, **44**(6), pp. 501–512.

[34] YAO, Y.-W., Z.-L. HOU, and Y.-Y. LIU (2007) "Transmission Frequency Properties of Elastic Waves along a Hetero-Phononic Crystal Waveguide," *Chinese Physics Letters Chinese Physics Letters*, **24**(2), p. 468.

[35] OLSSON-III, R. H. and I. EL-KADY (2009) "Microfabricated phononic crystal devices and applications," *Measurement Science and Technology*, **20**(1), p. 012002.

[36] IMAMURA, K. and S. TAMURA (2004) "Negative refraction of phonons and acoustic lensing effect of a crystalline slab," *Physical Review B*, **70**(17), p. 174308.

[37] ZHANG, X. and Z. LIU (2004) "Negative refraction of acoustic waves in two-dimensional phononic crystals," *Applied Physics Letters*, **85**(2), p. 341.

[38] CHEN, L.-S., C.-H. KUO, and Z. YE (2004) "Acoustic imaging and collimating by slabs of sonic crystals made from arrays of rigid cylinders in air," *Applied Physics Letters*, **85**(6), p. 1072.

[39] FENG, L., X.-P. LIU, Y.-B. CHEN, Z.-P. HUANG, Y.-W. MAO, Y.-F. CHEN, J. ZI, and Y.-Y. ZHU (2005) "Negative refraction of acoustic waves in two-dimensional sonic crystals," *Physical Review B*, **72**(3), p. 033108.

[40] SUKHOVICH, A., L. JING, and J. H. PAGE (2008) "Negative refraction and focusing of ultrasound in two-dimensional phononic crystals," *Physical Review B*, **77**(1), p. 014301.

[41] HÅKANSSON, A., F. CERVERA, and J. SÁNCHEZ-DEHESA (2005) "Sound focusing by flat acoustic lenses without negative refraction," *Applied Physics Letters*, **86**(5), p. 054102.

[42] CABALLERO, D., J. SÁNCHEZ-DEHESA, C. RUBIO, R. MÁRTINEZ-SALA, J. V. SÁNCHEZ-PÉREZ, F. MESEGUER, and J. LLINARES (1999) "Large two-dimensional sonic band gaps," *Physical Review E*, **60**(6), pp. R6316–R6319.

[43] GOFFAUX, C. and J. P. VIGNERON (2001) "Theoretical study of a tunable phononic band gap system," *Physical Review B*, **64**(7), p. 075118.

[44] KHELIF, A., P. A. DEYMIER, B. DJAFARI-ROUHANI, J. O. VASSEUR, and L. DOBRZYNSKI (2003) "Two-dimensional phononic crystal with tunable narrow pass band: Application to a waveguide with selective frequency," *Journal of Applied Physics*, **94**(3), p. 1308.

[45] HOU, Z., X. FU, and Y. LIU (2003) "Acoustic wave in a two-dimensional composite medium with anisotropic inclusions," *Physics Letters A*, **317**(1-2), pp. 127–134.

[46] YEH, J.-Y. (2007) "Control analysis of the tunable phononic crystal with electrorheological material," *Physica B: Condensed Matter*, **400**(1-2), pp. 137–144.

[47] Bertoldi, K. and M. C. Boyce (2008) "Mechanically triggered transformations of phononic band gaps in periodic elastomeric structures," *Physical Review B*, **77**(5), p. 052105.

[48] Yang, W.-P. and L.-W. Chen (2008) "The tunable acoustic band gaps of two-dimensional phononic crystals with a dielectric elastomer cylindrical actuator," *Smart Materials and Structures*, **17**(1), p. 015011.

[49] Yang, W.-P., L.-Y. Wu, and L.-W. Chen (2008) "Refractive and focusing behaviours of tunable sonic crystals with dielectric elastomer cylindrical actuators," *Journal of Physics D: Applied Physics*, **41**(13), p. 135408.

[50] Robillard, J.-F., O. B. Matar, J. O. Vasseur, P. A. Deymier, M. Stippinger, A.-C. Hladky-Hennion, Y. Pennec, and B. Djafari-Rouhani (2009) "Tunable magnetoelastic phononic crystals," *Applied Physics Letters*, **95**(12), p. 124104.

[51] Lin, S.-C. S. and T. J. Huang (2009) "Acoustic mirage in two-dimensional gradient-index phononic crystals," *Journal of Applied Physics*, **106**(5), p. 053529.

[52] Gómez-Reino, C., M. V. Perez, and C. Bao (2002) *Gradient-index Optics: Fundamentals and Applications*, Springer, Berlin.

[53] Medwin, H. and colleagues (2005) *Sounds in the Sea: From Ocean Acoustics to Acoustical Oceanography*, Cambridge University Press, Cambridge.

[54] Centeno, E., D. Cassagne, and J.-P. Albert (2006) "Mirage and superbending effect in two-dimensional graded photonic crystals," *Physical Review B*, **73**(23), p. 235119.

[55] Lin, S.-C. S., T. J. Huang, J.-H. Sun, and T.-T. Wu (2009) "Gradient-Index Phononic Crystals," *Physical Review B*, **79**(9), p. 094302.

[56] Akmansoy, E., E. Centeno, K. Vynck, D. Cassagne, and J.-M. Lourtioz (2008) "Graded photonic crystals curve the flow of light: An experimental demonstration by the mirage effect," *Applied Physics Letters*, **92**(13), p. 133501.

[57] Climente, A., D. Torrent, and J. Sánchez-Dehesa (2010) "Sound focusing by gradient index sonic lenses," *Applied Physics Letters*, **97**(10), p. 104103.

[58] Peng, S., Z. He, H. Jia, A. Zhang, C. Qiu, M. Ke, and Z. Liu (2010) "Acoustic far-field focusing effect for two-dimensional graded negative refractive-index sonic crystals," *Applied Physics Letters*, **96**(26), p. 263502.

[59] Lin, S.-C. S., B. R. Tittmann, J.-H. Sun, T.-T. Wu, and T. J. Huang (2009) "Acoustic beamwidth compressor using gradient-index phononic crystals," *Journal of Physics D: Applied Physics*, **42**(18), p. 185502.

[60] Wu, T.-T., Y.-T. Chen, J.-H. Sun, S.-C. S. Lin, and T. J. Huang (2011) "Focusing of the lowest antisymmetric Lamb wave in a gradient-index phononic crystal plate," *Applied Physics Letters*, **98**(17), p. 171911.

[61] Campbell, C. K. (1998) *Surface Acoustic Wave Devices for Mobile and Wireless Communications*, Academic, San Diego.

[62] Vasseur, J. O., A.-C. Hladky-Hennion, B. Djafari-Rouhani, F. Duval, B. Dubus, Y. Pennec, and P. A. Deymier (2007) "Waveguiding in two-dimensional piezoelectric phononic crystal plates," *Journal of Applied Physics*, **101**(11), p. 114904.

[63] Chen, J.-J., B. Bonello, and Z.-L. Hou (2008) "Plate-mode waves in phononic crystal thin slabs: Mode conversion," *Physical Review E*, **78**(3), p. 036609.

[64] Lin, S.-C. S., B. R. Tittmann, and T. J. Huang (2012) "Design of acoustic beam aperture modifier using gradient-index phononic crystals," *Journal of Applied Physics*, **111**(12), p. 123510.

[65] Lin, S.-C. S. and T. J. Huang (2011) "Tunable phononic crystals with anisotropic inclusions," *Physical Review B*, **83**(17), p. 174303.

[66] Kuang, W., Z. Hou, and Y. Liu (2004) "The effects of shapes and symmetries of scatterers on the phononic band gap in 2D phononic crystals," *Physics Letters A*, **332**(5-6), pp. 481–490.

[67] Yao, Y., Z. Hou, and Y. Liu (2007) "The two-dimensional phononic band gaps tuned by the position of the additional rod," *Physics Letters A*, **362**(5-6), pp. 494–499.

[68] Huang, Z.-G. and T.-T. Wu (2005) "Temperature effect on the bandgaps of surface and bulk acoustic waves in two-dimensional phononic crystals," *IEEE Trans. Ultrason., Ferroelect., Freq. Contr.*, **52**(3), pp. 365–370.

[69] Jim, K. L., C. W. Leung, S. T. Lau, S. H. Choy, and H. L. W. Chan (2009) "Thermal tuning of phononic bandstructure in ferroelectric ceramic/epoxy phononic crystal," *Applied Physics Letters*, **94**(19), p. 193501.

[70] Auld, B. (1990) *Acoustic fields and waves in solids*, vol. 1 of *Acoustic Fields and Waves in Solids*, R.E. Krieger.

[71] Kurt, H. and D. S. Citrin (2007) "Graded index photonic crystals," *Optics Express*, **15**(3), pp. 1240–1253.

[72] Juluri, B. K., S.-C. S. Lin, T. R. Walker, L. Jensen, and T. J. Huang (2009) "Propagation of Designer Surface Plasmons in Structured Conductor Surfaces with Parabolic Gradient Index," *Optics Express*, **17**(4), pp. 2997–3006.

[73] Goffaux, C., J. Sánchez-Dehesa, and P. Lambin (2004) "Comparison of the sound attenuation efficiency of locally resonant materials and elastic band-gap structures," *Physical Review B*, **70**(18), p. 184302.

Made in the USA
Middletown, DE
22 December 2017